Prairie I

Editor
Chandice M. Johnson, Jr.,
First Year English Committee
Dr. Muriel Brown
Dr. Robert O'Connor
Julie Bergman
Maureen Jonason
Helen Correll
North Dakota State University

McGraw-Hill, Inc.
College Custom Series

*New York St. Louis San Francisco Auckland Bogotá
Caracas Lisbon London Madrid Mexico Milan Montreal
New Delhi Paris San Juan Singapore Sydney Tokyo Toronto*

McGraw-Hill's **College Custom Series** consists of products that are produced from camera-ready copy. Peer review, class testing, and accuracy are primarily the responsibility of the author(s).

PRAIRIE PERSPECTIVES

1 2 3 4 5 6 7 8 9 0 MALMAL 9 0 9 8 7 6 5 4 3 2 1

ISBN 0-07-032678-9

Editor: Jon K. Earl

Cover Design: Kira Sholeen

Cover Photo Courtesy of Carrie Haak.

Printer/Binder: Malloy Lithographing, Inc.

This book is printed on recycled paper.

Table of Contents

Journal of Alexander Henry: 1800-1801 1

Lewis and Clark in North Dakota 4

Journals of Stephen H. Long and
James E. Colhoun: 1823 17

The Wonderful Ride 22

Homestead Days 35

An Urban Norwegian in the
Red River Valley: 1883-1885 75

Two Views of the Depression 101

Hardship 115

Velva, North Dakota: 1946 and 1964 116

Depot at Alice, Medora, Slant Village, Motel, and
Fort Berthold Reservation 122

A Place Unvisited 129

Winter Interlude 133

Buffalo Hunt 148

The Buffalo Coat 159

Acknowledgments 161

Preface

In *Not So Wild a Dream*, Eric Sevareid, who grew up in North Dakota, asks a series of rehetorical questions:

> North Dakota. Why have I not returned for so many years? Why have so few from the prairies never returned? Where is its written chapter in the long and varied American story? In distant cities when someone would ask, "Where are you from?" and I would answer: "North Dakota," they would merely nod politely and change the subject, having no point of common reference. They knew no one else from there. It was a large, rectangular blank spot in the nation's mind.

The selections in this anthology, which were written by individuals who have experienced North Dakota, have been chosen because they provide perspectives on North Dakota's written chapter in the American story and common reference points for those whom the state is a large, rectangular blank spot.

Acknowledgments

The present First-Year English Committee wishes to acknowledge its debt to the vision and efforts of earlier committees, who were responsible for creating the historical perspectives sections of the first and second editions of *Kaleidoscope 2: Emerging Perspectives on History and Social Justice*, which provided the basic format for this book. And while many and varied have been the contributions of our colleagues in North Dakota State University's English Department, note must be taken of the particular efforts of Mary Wallum, Eunice Johnston, Jane Kegel, and Tom Matchie, editors of the first edition of *Kaleidoscope 2*, and of Richard Shaw, editor of the second edition. Acknowledgement must also be made of the support given the committee by Dr. William Cosgrove, English Department Chair.

Journal of Alexander Henry: 1800-1801

Alexander Henry was a Canadian fur trader and one of the first white men to set up a trading post in North Dakota. He also was one of the few early explorers to keep a detailed journal filled with keen observations of daily life. In the entries printed below, Henry writes about his first post on the Red River near the mouth of the Park River in the winter of 1800-1801. His companion, Jean Baptiste Desmarais, is a French-Canadian interpreter. Crow, Charlo, Maymiutch, Liard, and Aceguemanche are Objibway Indians in Henry's party. As the journal suggests, the Objibway were very much afraid of the Sioux Indians who frequented the area.

Sunday, Sept. 7th. At nine o'clock sharp I sent off the canoes. The river makes a bend to the W. for some miles, when it resumes its southern direction. The river is deep, and at the forks, and the current is not very strong; no stones are to be seen. Desmarais and myself kept on ahead of the canoes. We saw buffalo in abundance, but did not disturb them, having plenty of fresh meat. At twelve o'clock we waited for the canoes; they arrived in three hours, when it began to rain, and we camped.

They had difficulty in coming up some shallow places, where the channels were narrow and difficult to find amongst the lumps of clay. Our situation not being proper for defense, and the rain coming on, I sent off the canoes, whilst Desmarais and myself went on ahead. We found a young bull with a broken leg. We killed him to examine the wound and found he had been shot with a ball in two places; this made us rather uneasy. For the first time I perceived that Desmarais feared the Sioux were at hand. However, we determined to say nothing about this affair to the men, who were getting very uneasy and wished me to pitch upon a place to build a fort. We came on about three miles and waited for the canoes. Here we saw another wounded bull, which increased our anxiety. The buffalo, however, appeared to be quiet. Whilst we waited for the canoes, several herds on the E. side came down to drink. We observed, also, on the opposite beach no fewer than seven bears drinking, all at the same time. The red deer were whistling in every direction, but our minds were not sufficiently at ease to enjoy our situation. A large white wolf, prowling rather imprudently, came within a few yards of us, and I shot him dead. When the canoes arrived the men swore they had seen horse tracks along the shore, but I did not believe them, as I supposed it to be buffalo tracks. . . .

Sept. 11th. I climbed up a tall oak, which I had trimmed for that purpose, at the entrance of the plain, from the top of which I had an extensive view of the country. Buffalo and red deer were everywhere in sight, passing to and fro. The weather being perfectly serene, I could distinguish the Hair hills on the W., though they were scarcely perceptible—nothing more than a blue stripe, running N. and S. The interval is a level meadow, with nothing to attract the eye but the winding course of the Park river, whose wood is lost to the sight long before it reaches the hills. The distance may be between 12 and 15 leagues.

At ten o'clock Crow arrived on the only horse the Indians have. He informed me that last summer, when they went to war, they passed along the Hair hills, and that exactly opposite us, on the Middle branch of Park river, and at the foot of the hills, they fell upon a large camp of the enemy, where it appeared they had been tented a long time, and had dried vast quantities of buffalo flesh, but had decamped about 20 days before. This afternoon Charlo and family arrived on foot. He wanted liquor for four bearskins, but I would give him none until all the Indians arrived, when I proposed to sell them liquor for what few skins and provisions they might have; as giving them rum for nothing was out of the question. I had already given them as much as they deserved, and for the future, if they wanted liquor, they must expect to pay for it.

This afternoon my storehouse was finished; I immediately put everything into it, and a padlock on the door. I now desired each man to cut 50 oak stockades 12 feet long and carry them to the spot where I proposed erecting the fort. Maymiutch, Liard, and Aceguemanche arrived and camped; they had killed 12 bears. Crow, who had gone hunting, returned, having killed two bears. I shot a wolf that was passing on the opposite side of the river, and killed him dead. My double-barreled gun is an excellent piece. Buffaloes come down to drink, both day and night, near our camp; we seldom molest them, but allow them to return in quiet. The Indians this evening loaded our stage with bears' fat and choice meat. This would have been a glorious time for my men, had not dread of the Sioux deprived them of their appetite and made them only anxious to finish the fort.

Jan. 14th. At daybreak I was awakened by the bellowing of buffaloes. I got up, and was astonished when I climbed into the S. W. bastion. On my right the plains were black, and appeared as if in motion, S. to N. Opposite the fort the ice was covered; and on my left, to the utmost extent of the reach below us, the river was covered with buffalo moving northward. Our dogs were confined within the fort, which allowed the buffalo to pass within a few paces. I dressed and climbed my oak for a better view. I had seen almost incredible numbers of buffalo in the fall, but nothing in comparison to what I now beheld. The ground was covered at every point of the compass, as far as the eye could reach, and every animal was in motion. All hands soon attacked them with a tremendous running fire, which put them to a quicker pace, but had no effect in altering their course. The first roads beaten in the snow were followed by those in the rear. They passed at full speed until about nine o'clock, when their numbers decreased and they kept further off in the plain. There was about 15 inches of snow on a level, in some places drifted in great banks. Notwithstanding the buffalo were so numerous, and 12 guns were employed, we killed only three cows and one old bull, but must have wounded a great number.

Wednesday, Apr. 1st. The river is clear of ice, but drowned buffalo continue to drift by in entire herds. Several are lodged on the banks near the fort. The women cut up some of the fattest for their own use; the flesh appeared to be fresh and good. It is really astonishing what vast numbers have perished; they formed one continuous line in the current for two days and nights. One of my men found a herd that had fallen through the ice in Park river and all been drowned; they were sticking in the ice, which had not yet moved in that part. The women had excellent sport in raising the back fat and tongues. On the 5th, the plains having been clear of snow and dry, we had two feet of snow, and the river nearly froze over again. We brought our baggage into the fort.

Lewis and Clark in North Dakota

The Lewis and Clark Expedition was in what is now North Dakota from October 13, 1804 until April 27, 1805, and from August 3 until August 20, 1806. Shortly after arriving in the region in October, 1804, the expedition went into winter camp near the present location of the city of Mandan, North Dakota. The journal entries reproduced here show North Dakota from the perspective of some of its earliest visitors. (Note: Entries are from the Nicholas Biddle edition of the journals.)

15 October 1804 [Location: east bank of the Missouri, three miles north of present site of Ft. Yates, N.D.] We stopped at three miles on the north a little above a camp of Ricaras who are hunting, where we were visited by about thirty Indians. They came over in their skin canoes, bringing us meat, for which we returned them beads and fishhooks. About a mile higher we found another encampment of Ricaras on the south, consisting of eight lodges: here we again ate and exchanged a few presents. As we went we discerned numbers of other Indians on both sides of the river; and at about nine miles we came to a creek on the south, where we saw many high hills resembling a house with slanting roof; and a little below the creek an old village of the Sharha or Chayenne Indians. The morning had been cloudy, but the evening became pleasant, the wind from the northeast, and at sunset we halted, after coming ten miles over several sandbars and points, above a camp of ten Ricara lodges on the north side. We visited their camp, and smoked and eat with several of them; they all appeared kind and pleased with our attentions, and the fair sex received our men with more than hospitality. York was here again an object of astonishment; the children would follow him constantly, and if he chanced to turn towards them, run with great terror. The country of to-day is generally low and covered with timber on both sides, though in the morning we passed some barren hills on the south.

17 October 1804 [Location: west bank of the Missouri, about three miles south of the Cannonball River.] The weather was pleasant: we passed a low ground covered with small timber on the south, and barren hills on the north which came close to the river; the wind from the northwest then became so strong that we could not move after ten o'clock, until late in the afternoon, when we were forced to use the towline, and we therefore made only six miles. We all went out hunting and examining the country. The goats, of which we see large flocks coming to the north bank of the river, spend the summer, says Mr. Gravelines, in the plains east of the Missouri, and at the present season are returning to the Black mountains, where they subsist on leaves and shrubbery during the winter, and resume their migrations in the spring. We also saw buffalo, elk, and deer, and a number of snakes; a beaver house too was seen, and we caught a whip-poor-will of a small and uncommon kind: the leaves are fast falling; the river wider than usual and full of sandbars; and on the sides of the hills are large stones, and some rock of a brownish colour in the southern bend below us. Our latitude by observation was 45° 23' 57".

20 October 1804 [Location: west bank of the Missouri, where Ft. Lincoln park is now located. Village mentioned is now known as "Slant Village." See Lyons' poem.] We proceeded early with a southeast wind, which continued high all day, and came to a creek on the north at two miles distance, twenty yards wide. At eight miles we reached the lower point of an island in the middle of the river, though there is no current on the south. This island is covered with willows and extends about two miles, there being a small creek coming in from the south at its lower extremity. After making twelve miles we encamped on the south, at the upper part of a bluff containing stone-coal of an inferior quality; immediately below this bluff and on the declivity of a hill, are the remains of a village covering six or eight acres, formerly occupied by the Mandans, who, says our Ricara chief, once lived in a number of villages on each side of the river, till the Sioux forced them forty miles higher; whence, after a few years residence, they moved to their present position. The country through which we passed has wider bottoms and more timber than those we have been accustomed to see, the hills rising at a distance and by gradual ascents. We have seen great numbers of elk, deer, goats, and buffalo, and the usual attendants of these last, the wolves, who follow their movements and feed upon those who die by accident, or who are too poor to keep pace with the herd; we also wounded a white bear, and saw some fresh tracks of those animals which are twice as large as the track of a man.

21 October 1804 [Location: near the mouth of the Heart River.] Last night the weather was cold, the wind high from the northeast, and the rain which fell froze on the ground. At daylight it began to snow, and continued till the afternoon, when it remained cloudy and the ground was covered with snow. We however, set out early, and just above our camp came to a creek on the south, called Chisshetaw, about thirty yards wide and with a considerable quantity of water. Our Ricara chief tells us, that at some distance up this river is situated a large rock which is held in great veneration, and visited by parties who go to consult it as to their own or their nation's destinies, all of which they discern in some sort of figures or paintings with which it is covered. About two miles off from the mouth of the river the party on shore saw another of the objects of Ricara superstition: it is a large oak tree, standing alone in the open prairie, and as it alone has withstood the fire which has consumed every thing around, the Indians naturally ascribe to it extraordinary powers. One of their ceremonies is to make a hole in the skin of their necks through which a string is passed and the other end tied to the body of the tree; and after remaining in this way for some time they think they become braver. At two miles from our encampment we came to the ruins of a second Mandan village, which was in existence at the same time with that just mentioned. It is situated on the north at the foot of a hill in a beautiful and extensive plain, which is now covered with herds of buffalo: nearly opposite are remains of a third village on the south of the Missouri; and there is another also about two miles further on the north, a little off the river. At the distance of seven miles we encamped on the south, and spent a cold night. We procured a buffalo and an otter only. The river is wide and the sandbars numerous, and a low island near our encampment.

22 October 1804 [Location: site of present day Mandan, N.D. Journal entries cover period from the explorers' arrival in the Mandan area October 22 to their selection of a spot for winter camp November 2.] In the morning we passed an old Mandan village on the south, near our camp; at four miles another on the same side. About seven o'clock we came to at a camp of eleven Sioux of the Teton tribe, who are almost perfectly naked, having only a piece of skin or cloth round the middle, though we are suffering from the cold. From their appearance, which is warlike, and from their giving two different accounts of themselves, we believe that they are either going to or returning from the Mandans, to which nations the Sioux frequently make excursions to steal horses. As their conduct displeased us, we gave them nothing. At six we reached an island about one mile in length, at the head of which is a Mandan village on the north in ruins, and two miles beyond a bad sandbar. At eight miles are remains of another Mandan village on the south; and at twelve miles encamped on the south. The hunters brought in a buffalo bull, and mentioned that of about three hundred which they had seen, there was not a single female. The beaver is here in plenty, and the two Frenchmen who are returning with us catch several every night.

These villages which are nine in number are scattered along each side of the river within a space of twenty miles; almost all that remains of them is the wall which surrounded them, the fallen heaps of earth which covered the houses, and occasionally human skulls and the teeth and bones of men, and different animals, which are scattered on the surface of the ground.

23 October . The weather was cloudy and we had some snow; we soon arrived at five lodges where the two Frenchmen had been robbed, but the Indians had left it lately as we found the fires still burning. The country consists as usual of timbered low grounds, with grapes, rushes, and great quantities of a small red acid fruit, known among the Indians by a name signifying rabbitberries, and called by the French *graisse de buffle* or buffalo fat. The river too, is obstructed by many sandbars. At twelve miles we passed an old village on the north, which was the former residence of the Ahnahaways who now live between the Mandans and Minnetarees. After making thirteen miles we encamped on the south.

24 October. The day was again dark and it snowed a little in the morning. At three miles we came to a point on the south, where the river by forcing a channel across a former bend has formed a large island on the north. On this island we found one of the grand chiefs of the Mandans, who with five lodges was on a hunting excursion. He met his enemy the Ricara chief, with great ceremony and apparent cordiality, and smoked with him. After visiting his lodges, the grand chief and his brother came on board our boat for a short time; we then proceeded and encamped on the north, at seven miles from our last night's station and below the old village of the Mandans and Ricaras. Here four Mandans came down from a camp above, and our Ricara chief returned with them to their camp, from which we augur favourably of their pacific views towards each other. The land is low and beautiful, and covered with oak and cottonwood, but has been too recently hunted to afford much game.

25 October. The morning was cold and the wind gentle from the southeast: at three miles we passed a handsome high prairie on the south, and on an eminence about forty feet above the water and extending back for several miles in a beautiful plain, was situated an old village of the Mandan nation which has been deserted for many years. A short distance above it, on the continuation of the same rising ground are two old villages of Ricaras, one on the top of the hill, the other in the level plain, which have been deserted only five years ago. Above these villages is an extensive low ground for several miles, in which are situated, at three or four miles from the Ricara villages, three old villages of Mandans near together. Here the Mandans lived when the Ricaras came to them for protection, and from this they moved to their present situation above. In the low ground the squaws raised their corn, and the timber, of which there was little near the villages, was supplied from the opposite side of the river, where it was and still is abundant.

As we proceeded several parties of Mandans both on foot and horseback came along the river to view us, and were very desirous that we should land

and talk to them: this we could not do on account of the sandbreaks on the shore, but we sent our Ricara chief to them in a periogue. The wind too having shifted to the southwest and being very high it required all our precautions on board, for the river was full of sandbars which made it very difficult to find the channel. We got around several times, and passed a very bad point of rocks, after which we encamped on a sandpoint to the north, above a handsome plain covered with timber, and opposite to a high hill on the south side at the distance of eleven miles. Here we were joined by our Ricara chief, who brought an Indian to the camp where he remained all night.

26 October. We set out early with a southwest wind, and after putting the Ricara chief on shore to join the Mandans who were in great numbers along it, we proceeded to the camp of the grand chiefs four miles distant. Here we met a Mr. M'Cracken one of the northwest or Hudson Bay company, who arrived with another person about nine days ago to trade for horses and buffalo robes. Two of the chiefs came on board with some of their household furniture, such as earthen pots and a little corn and went on with us; the rest of the Indians following on shore. At one mile beyond the camp we passed a small creek, and at three more a bluff of coal of an inferior quality on the south. After making eleven miles we reached an old field where the Mandans had cultivated grain last summer, and encamped for the night on the south side, about half a mile below the first village of the Mandans. In the morning we had a willow low ground on the south and highland on the north, which occasionally varied in the course of the day. There is but little wood on this part of the river, which is here subdivided into many channels and obstructed by sandbars. As soon as we arrived a crowd of men, women, and children came down to see us. Captain Lewis returned with the principal chiefs to the village, while the others remained with us during the evening; the object which seemed to surprise them most, was a cornmill fixed to the boat which we had occasion to use, and delighted them by the ease with which it reduced the grain to powder. Among others who visited us was the son of the grand chief of the Mandans, who had his two little fingers cut off at the second joints. On inquiring into this accident, we found that it was customary to express grief for the death of relations by some corporeal suffering, and that the usual mode was to lose two joints of the little fingers, or sometimes the other fingers. The wind blew very cold in the evening from the southwest. Two of the party are affected with rheumatic complaints.

27 October. At an early hour we proceeded and anchored off the village. Captain Clark went on shore, and after smoking a pipe with the chiefs, was desired to remain and eat with them. He declined on account of his being unwell; but his refusal gave great offence to the Indians, who considered it disrespectful not to eat when invited, till the cause was explained to their satisfaction. We sent them some tobacco, and then proceeded to the second village on the north, passing by a bank containing coal, and a second village, and encamped at four miles on the north, opposite to a village of Ahnahaways. We here met with a Frenchman, named Jesseaume, who lives among the Indians with his wife and children, and who we take as an interpreter. The Indians had flocked to the bank to see us as we passed, and they visited in great numbers the camp, where some of them remained all night. We sent in the evening three young Indians with a present of tobacco for the chiefs of the three upper villages, inviting them to come down in the morning to a council with us. Accordingly the next day,

28 October. We were joined by many of the Minnetarees and Ahnahaways from above, but the wind was so violent from the southwest that the chiefs of the lower villages could not come up; and the council was deferred till to-morrow. In the mean while we entertained our visitors by showing them what was new to them in the boat; all which, as well our black servant, they called Great Medicine, the meaning of which we afterwards learnt. We also consulted the grand chief of the Mandans, Black Cat, and Mr. Jesseaume, as to be the names, characters, &c., the chiefs with whom we are to hold the council. In the course of the day we received several presents from the women, consisting of corn, boiled hominy, and garden stuffs: in our turn we gratified the wife of the great chief with a gift of a glazed earthen jar. Our hunter brought us two beaver. In the afternoon we sent the Minnetaree chiefs to smoke for us with the great chief of the Mandans, and told them we would speak in the morning.

Finding that we shall be obliged to pass the winter at this place, we went up the river about one and a half miles to-day, with a view of finding a convenient spot for a fort, but the timber was too scarce and small for our purposes.

29 October. The morning was fine and we prepared our presents and speech for the council. After breakfast we were visited by an old chief of the Ahnahaways, who finding himself growing old and weak had transferred his power to his son, who is now at war against the Shoshonees. At ten o'clock the chiefs were all assembled under an awning of our sails, stretched so as to exclude the wind which had become high; that the impression might be more forcible, the men were all paraded, and the council opened by a discharge from the swivel of the boat. We then delivered a speech, which like those we had already made intermingled advice with assurances of friendship and trade: while we were speaking the old Ahnahaway chief grew very restless, and observed that he could not wait long as his camp was exposed to the hostilities of the Shoshonees; but he was instantly rebuked with great dignity by one of the chiefs for this violation of decorum at such a moment, and remained quiet during the rest of the council. Towards the end of our speech we introduced the subject of our Ricara chief, with whom we recommended a firm peace: to this they seemed well disposed, and all smoked with him very amicably. We all mentioned the goods which had been taken from the Frenchmen, and expressed a wish that they should be restored. This being over, we proceeded to distribute the presents with great ceremony: one chief of each town was acknowledged by a gift of a flag, a medal with the likeness of the president of the United States, a uniform coat, hat and feather: to the second chiefs we gave a medal representing some domestic animals, and a loom for weaving; to the third chiefs medals with the impressions of a farmer sowing grain. A variety of other presents were distributed, but none seemed to give them more satisfaction than an iron corn mill which we gave to the Mandans.

The chiefs who were made to-day are: Shahaka or Big White, a first chief, and Kagohami or Little Raven, a second chief of the lower village of the Mandans, called Matootonha: the other chiefs of an inferior quality who were recommended were, 1. Ohheenaw, or Big Man, a Chayenne taken prisoner by the Mandans who adopted him, and he now enjoys great consideration among the tribe. 2. Shotahawrora, or Coal, of the second Mandan village which is called Rooptahee. We made Poscopsahe, or Black Cat, the first chief of the village, and the grand chief of the whole Mandan nation: his second chief is Kagonomokshe, or Raven man Chief; inferior chiefs of this village were, Tawnuheo, and Bellahsara, of which we did not learn the translation.

In the third village which is called Mahawha, and where the Arwacahwas reside, we made one first chief, Tetuckopinreha, or White Buffalo robe unfolded, and recognized two of an inferior order: Minnissurraree, or Neighing Horse, and Locongotiha, or Old woman at a distance.

Of the fourth village where the Minnetarees live, and which is called Metaharta, we made a first chief, Ompsehara, or Black Moccasin: a second chief, Ohhaw, or Little Fox. Other distinguished chiefs of this village were, Mahnotah, or Big Thief, a man whom we did not see as he is out fighting, and was killed soon after; and Mahserassa, or Tail of the Calumet Bird. In the fifth village we made a first chief Eapanopa, or Red Shield; a second chief Wankerassa, or Two Tailed Calumet Bird, both young chiefs; other persons of distinction are, Shahakohopinnee, or Little Wolf's Medicine; Ahrattanamockshe, or Wolfman chief, who is now at war, and is the son of the old chief we have mentioned, whose name is Caltahcota, or Cherry on a Bush.

The presents intended for the grand chief of the Minnetarees, who was not at the council, were sent to him by the old chief Caltahcota; and we delivered to a young chief those intended for the chief of the lower villages.

The council was concluded by a shot from our swivel, and after firing the airgun for their amusement, they retired to deliberate on the answer which they are to give to-morrow.

In the evening the prairie took fire, either by accident or design, and burned with great fury, the whole plain being enveloped in flames: so rapid was its progress that a man and a woman were burnt to death before they could reach a place of safety; another man with his wife and child were much burnt, and several other persons narrowly escaped destruction. Among the rest a boy of the half white breed escaped unhurt in the midst of the flames; his safety was ascribed to the great medicine spirit, who had preserved him on account of his being white. But a much more natural cause was the presence of mind of his mother, who seeing no hopes of carrying off her son, threw him on the ground, and covering him with the fresh hide of a buffalo, escaped herself from the flames; as soon as the fire had passed, she returned and found him untouched, the skin having prevented the flame from reaching the grass on which he lay.

30 October. We were this morning visited by two persons from the lower village, one the Big White the chief of the village, the other the Chayenne called the Big Man; they had been hunting, and did not return yesterday early enough to attend the council. At their request we repeated part of our speech of yesterday, and put the medal round the neck of the chief. Captain Clark took a periogue and went up the river in search of a good wintering place, and returned after going seven miles to the lower point of an island on the north side, about one mile in length; he found the banks on the north side high, with coal occasionally, and the country fine on all sides; but the want of wood and the scarcity of game up the river, induced us to decide on fixing ourselves lower down during the winter. In the evening our men danced among themselves to the great amusement of the Indians.

31 October. A second chief arrived this morning with an invitation from the grand chief of the Mandans, to come to his village where he wished to present some corn to us and to speak with us. Captain Clark walked down to his village; he was first seated with great ceremony on a robe by the side of the chief, who then threw over his shoulder another robe handsomely ornamented. The pipe was then smoked with several of the old men who were seated around the chief; after some time he began his discourse, by observing that he believed what we had told him, and that they should soon enjoy peace, which would gratify him as well as his people, because they could then hunt without fear of being attacked, and the women might work in the fields without looking every moment for the enemy, and at night put off their moccasins, a phrase by which is conveyed the idea of security when the women could undress at night without fear of attack. As to the Ricaras, he continued, in order to show you that we wish peace with all men, that chief, pointing to his second chief, will go with some warriors back to the Ricaras with their chief now here and smoke with that nation. When we heard of your coming all the nations around returned from their hunting to see you, in hopes of receiving large presents; all are disappointed and some discontented; for his part he was not much so, though his village was. He added that he would go and see his great father the president. Two of the steel traps stolen from the Frenchmen were then laid before captain Clark, and the women brought about twelve bushels of corn. After the chief had finished, captain Clark made an answer to the speech and then returned to the boat, where he found the chief of the third village and Kagohami (the Little Raven) who smoked and talked about an hour. After they left the boat the grand chief of the Mandans came dressed in the clothes we had given him, with his two children, and begged to see the men dance, in which they willingly gratified him.

1 November. Mr. M'Cracken, the trader whom we found here, set out to-day on his return to the British fort and factory on the Assiniboin river, about one hundred and fifty miles from this place. He took a letter from captain Lewis to the northwest company, inclosing a copy of the passport granted by the British minister in the United States. At ten o'clock the chiefs of the lower village arrived they requested that we would call at their village for some corn, that they were willing to make peace with the Ricaras, that they had never provoked the war between them, but as the Ricaras had killed some of their chiefs they had retaliated on them; that they had killed them like birds, till they were tired of killing them, so that they would send a chief and some warriors to smoke with them. In the evening we dropped down to the lower village where captain Lewis went on shore, and captain Clark proceeded to a point of wood on the north side.

2 November. He therefore went up to the village where eleven bushels of corn were presented to him. In the meantime captain Clark went down with the boats three miles, and having found a good position where there was plenty of timber, encamped and began to fell trees to build our huts. Our Ricara chief set out with one Mandan chief and several Minnetaree and Mandan warriors; the wind was from the southeast, and the weather being fine a crowd of Indians came down to visit us.

7 April 1805 [Camp broken, and travel resumed.] He [the interpreter to the villages] returned with a Ricara chief and three of his nation. The chief, whose name is Kagowheta, or Brave Raven, brought a letter from Mr. Tabeau, mentioning the wish of the grand chiefs of the Ricaras to visit the president, and requesting permission for himself and four men to join our boat when it descends; to which we consented, as it will then be manned with fifteen hands and be able to defend itself against the Sioux. After presenting the letter, he told us that he was sent with ten warriors by his nation to arrange their settling near the Mandans and Minnetarees, whom they wished to join; that he considered all the neighbouring nations friendly except the Sioux, whose persecution they would no longer withstand, and whom they hoped to repel by uniting with the tribes in this quarter: he added that the Ricaras intended to follow our advice and live in peace with all nations, and requested that we would speak in their favour to the Assiniboin Indians. This we willingly promised to do, and assured them that their great father would protect them and no longer suffer the Sioux to have good guns, or to injure his dutiful children. We then gave him a small medal, a certificate of his good conduct, a carrot of tobacco, and some wampum, with which he departed for the Mandan village well satisfied with his reception. Having made all our arrangements, we left the fort about five o'clock in the afternoon. The party now consisted of thirty-two persons. Besides ourselves were sergeants John Ordway, Nathaniel Pryor, and Patrick Gass: the privates were William Bratton, John Colter, John Collins, Peter Cruzatte, Robert Frazier, Reuben Fields, Joseph Fields, George Gibson, Silas Goodrich, Hugh Hall, Thomas P. Howard, Baptiste Lapage, Francis Labiche, Hugh M'Neal, John Potts, John Shields, George Shannon, John B. Thompson, William Werner, Alexander Willard, Richard Windsor, Joseph Whitehouse, Peter Wiser, and captain Clark's black servant York. The two interpreters, were George Drewyer and Toussaint Chaboneau. The wife of Chaboneau also accompanied us with her young child, and we hope may be useful as an interpreter among the Snake Indians. She was herself one of that tribe, but having been taken in war by the Minnetarees, by whom she was sold as a slave to Chaboneau, who brought her up and afterwards married her. One of the Mandans likewise embarked with us, in order to go the Snake Indians and obtain a peace with them for his countrymen. All this party with the baggage was stowed in six small canoes and two large periogues. We left the fort with fair pleasant weather though the northwest wind was high, and after making about four miles encamped on the north side of the Missouri, nearly opposite the first Mandan village. At the same time that we took our departure our barge manned with seven soldiers, two Frenchmen, and Mr. Gravelines as pilot, sailed for the United States loaded with our presents and despatches.

23 April 1805 [Location: present Williams County, near the mouth of Painted Woods Creek.] A clear and pleasant morning, but at nine o'clock the wind became so high that the boats were in danger of upsetting; we therefore were forced to stop at a place of safety till about five in the afternoon, when the wind being lower we proceeded and encamped on the north at the distance of thirteen and a half miles: the party on shore brought us a buffalo calf and three blacktailed deer: the sand on the river has the same appearances as usual, except that the quantity of wood increases.

24 April [Location: vicinity of site of Ft. Union from April 24 through April 26.] The wind blew so high during the whole day that we were unable to move; such indeed was its violence, that although we were sheltered by high timber the waves wet many articles in the boats: the hunters went out and returned with four deer, two elk, and some young wolves of the small kind. The party are very much afflicted with sore eyes, which we presume are occasioned by the vast quantities of sand which are driven from the sandbars in such clouds as often to hide from us the view of the opposite bank. The particles of this sand are so fine and light that it floats for miles in the air like a column of thick smoke, and is so penetrating that nothing can be kept free from it, and we are compelled to eat, drink, and breathe it very copiously. To the same cause we attribute the disorder of one of our watches, although her cases are double and tight; since without any defect in its works, that we can discover, it will not run for more than a few minutes without stopping.

25 April. The wind moderated this morning, but was still high; we therefore set out early, the water being so cold that the water froze on the oars as we rowed, and about ten o'clock the wind increased so much that we were obliged to stop. This detention from the wind and the reports from our hunters of the crookedness of the river, induced us to believe that we were at no great distance from the Yellowstone river. In order therefore to prevent delay as much as possible, captain Lewis determined to go on by land in search of that river and make the necessary observations, so as to be enable to proceed on immediately after the boats should join him: he therefore landed about eleven o'clock on the south side, accompanied by four men; the boats were prevented from going until five in the afternoon, when they went on a few miles further and encamped for the night at the distance of fourteen and a half miles.

26 April. We continued our voyage in the morning and by twelve o'clock encamped at eight miles distance, at the junction of the Missouri and Yellowstone rivers; where we were soon joined by captain Lewis.

On leaving us yesterday he pursued his route along the foot of the hills, which he ascended at the distance of eight miles; from these the wide plains watered by the Missouri and the Yellowstone spread themselves before the eye, occasionally varied with the wood of the banks, enlivened by the irregular windings of the two rivers, and animated by vast herds of buffalo, deer, elk, and antelope. The confluence of the two rivers was concealed by the wood, but the Yellowstone itself was only two miles distant to the south. He therefore descended the hills and encamped on the bank of the river, having killed as he crossed the plain four buffaloes the deer alone are shy and retired to the woods, but the elk, antelope, and buffalo suffered him to approach them without alarm, and often followed him quietly for some distance. This morning he sent a man up the river to examine it, while he proceeded down to the junction: the ground on the lower side of the Yellowstone near its mouth, is flat, and for about a mile seems to be subject to inundation, while that at the point of junction, as well as on the opposite side of the Missouri, is at the usual height of ten or eighteen feet above the water, and therefore not overflown. There is more timber in the neighbourhood of this place, and on the Missouri, as far below as the Whiteearth river, than on any other part of the Missouri on this side of the Chayenne: the timber consists principally of cottonwood with some small elm, ash, and box alder. On the sandbars and along the margin of the river grows the small-leafed willow; in the low grounds adjoining are scattered rosebushes three or four feet high, the redberry, serviceberry and redwood. The higher plains are either immediately on the river, in which case they are generally timbered, and have an undergrowth like that of the low grounds, with the addition of the broad-leafed willow, gooseberry, chokecherry, purple currant, and honeysuckle: or they are between the low grounds and the hills, and for the most part without wood or any thing except large quantities of wild hyssop; this plant rises about two feet high and like the willow of the sandbars is a favourite food of the buffalo, elk, deer, grouse, porcupine, hare, and rabbit. This river which had been known to the French as the Roche jaune, or as we have called it the Yellowstone, rises according to Indian information in the Rocky mountains; its sources are near those of the Missouri and the Platte, and it may be navigated in canoes almost to its head. It runs first through a mountainous country, but in many parts fertile and well timbered; it then waters a rich delightful land, broken into vallies and meadows and well supplied with wood and water till it reaches near the Missouri open meadows and low grounds, sufficiently timbered on its borders. In the upper country its course is represented as very rapid, but during the two last and largest portions, its current is much more gentle than that of the Missouri, which it resembles also in being turbid though with less sediment. The man who was sent up the river, reported in the evening that he had gone about eight miles, that during that distance the river winds on both sides of a plain four or five miles wide, that the current was gentle and much obstructed by sandbars, that at five miles he had met with a large timbered island, three miles beyond which a creek falls in on the S. E. above a high bluff, in which are several strata of coal. The country as far as he could discern, resembled that of the Missouri, and in the plain he met several of the bighorn animals, but they were too shy to be obtained. The bed of the Yellowstone, as we observed it near the mouth, is composed of sand and mud, without a stone of any kind. Just above the confluence we measured the two rivers, and found the bed of the Missouri five hundred and twenty yards wide, the water occupying only three hundred and thirty, and the channel deep: while the Yellowstone, including its sandbar, occupied eight hundred and fifty-eight yards, with two hundred and ninety-seven yards of water: the deepest part of the channel is twelve feet, but the river is now falling and seems to be nearly at its summer height.

Journals of Stephen H. Long and James E. Colhoun: 1823

On April 30, 1823, Major Stephen H. Long, of the United States Army Topographical Engineers, left Philadelphia, Pennsylvania, on a six-month expedition. Stopping near the present city of Minneapolis to replenish his supplies at Fort St. Anthony, he continued along the St. Peter's River (now the Minnesota) and up the Red River to Lake Winnipeg before heading east along the Great Lakes and back to Philadelphia. His mission was to provide a topographical description of his route, including latitude and longitude of all important landmarks, to mark the boundary between Canada and the United States at the 49th parallel near Pembina, North Dakota, and to gather information about the natural resources and the Indian tribes inhabiting the area.

When he left Ft. St. Anthony, near Minneapolis, on July 9,1823, Major Long split his party into two groups, one traveling on horseback along the river, the other traveling up the Minnesota and Red Rivers by flat-bottom skiff and canoe. The land group included James E. Colhoun; Thomas Say, a geologist; William Snelling, a volunteer guide; Giacomo Beltrami, an Italian traveler; Louis Demarais, a guide and Chippewah interpreter; Lieutenant St. Clair Denny; and nine soldiers. The water group, which was carrying most of the supplies, included Major Long; William H. Keating, a geologist; Samuel Seymour, a landscape painter; Joseph Renville, a guide and Sioux interpreter; and twelve soldiers under the command of a corporal. Joseph Jeffryes, a fur trader for the Columbia Fur Company, joined the expedition at Lake Travers on July 26. Because the two groups were frequently out of contact with each other, a clear picture of the expedition will emerge only from reading the journals of both Long and Colhoun.

The Journal of Stephen H. Long, July 28, 1823

Monday, July 28, 1823. Started at half past 6 a.m. and proceeded down the [Bois de] Sioux river to Red river, which we crossed about 5 miles below its confluence with the former. On crossing the river an accident happened which required considerable time to repair. The cart loaded with our dry meat overturned as it was descending to the water thro' the carelessness and obstinacy of its driver, and the contents of the cart precipitated into the water. We accordingly halted on the east shore of the river to dry the meat.

Red river has its source in a Lake about 5 miles wide and 24 long called Otter-tail Lake, which is situated in a N easterly direction from the Trading establishment at Lake Travers at the distance of about 150 miles. Near the source of this river are high lands, visible at our noon day encampment, denominated the Leaf mountains.

After drying our meat, which we were able to do pretty effectually, we resumed our march and travelled downward on the east side of the river. Innumerable buffaloes thronged the prairie and a large herd of elk, consisting probably of more than 100, presented themselves and were pursued by three gentlemen of the party, who killed one of them. This success however was likely to bring upon us serious evil. For while three of the party, including our guide & Interpreter [*Jeffryes*], were employed in this enterprise, the main party encountered a group of about 35 indians of the band called the Gens de la feuille, notorious for their rudeness and barbarity, whose deportment towards us was by no means destitute of these attributes.

They attempted to stop us on our march and demand presents, insisted upon our encamping at a place they would choose, but to no purpose but that of making us more determined in the course proper for us to take. We encamped about 5 miles above them thinking our horses, baggage &c. would be more secure from their depredations. We were very apprehensive that our companions who had engaged in the chase would be illy treated by them, but fortunately they arrived safely at camp at a late hour in the evening. Encamped at sunset. Dist. 22½ miles.

The Journal of James E. Colhoun, July 28, 1823

Monday, July 28,1823. Proceeded 8 miles and forded Red River, where it is 25 yards wide about 5 miles by land below the mouth of Swan River [Bois de Sioux]. In crossing the cart containing all our dried buffalo meat upset & its contents fell into the water. The accident detained us for some time, for it was necessary to expose to the sun the meat that had been wet. From where we crossed, Otter-tail Lake, the source of Red River, bears northeasterly distant 70 or 80 miles. One of our Canadians who has seen it says that it was about 24 [9] miles long from North to South & 4½ [3] wide. Mr. Jeffries describes the country lying in that direction as being full of small lakes & "Islands of Woods."

A Bald Eagle flew over our heads & alighted on her nest in our sight. Travelled over level prairie: no trees to be seen except a strip of woods along the Red River & the forest on the Montagnes des Feuilles [*Leaf Mountains]* a great distance to the Northeastward; the base of which range, that is, their southern [*western*] extreme, is washed by the [*Otter Tail*] River. We frequently see wolves [*coyotes*], & [the] horizon is occupied sometimes for several miles by a single "band" of buffalo. Bulls are sprinkled over the plain; they often advance within musket shot to look at us. Some of the party attribute this apparent boldness to the long hair hanging over their eyes & obstructing their vision, so as to hinder them from discovering, unless very near, that the moving crowd is not a band of cows. Others suppose them to have a degree of fearlessness in the rutting season.

Early in the afternoon two of the party besides myself went in pursuit of a large herd of Elk. By creeping on our hands & knees a quarter of a mile leading our horses, we were enabled to approach within 80 yards. We then fired & one of the herd, a female, fell. As had been previously agreed, having the best horse & a pair of pistols, I mounted and chaced for more than a mile. I got up with them in half that distance, but my horse was so much afraid of them that the severest spurring did not urge him near enough: I flashed one pistol & discharged the other without effect. In the pursuit those in the rear repeatedly stopped to look at me. When in herds, it is not difficult to overtake them; it is otherwise when they are alone; however, they are said to be short-winded. The Elk is called in this country the Red Deer; the Virginia [*white-tailed*] Deer, the species common to the Southern States, is here termed the Jumping Deer. This last kind is said never to be found North of lake Travers. It is on account of creeping through the grass, so often to be done, that every body in this country has a leather cover for his piece, which must be dragged along: the cover protects the gun from the dew & secures the hunter from the danger of its being accidentally cocked.

We were occupied until sunset cutting up the carcase, & it was late in the night be-fore we reached the camp on the bank of Red River, or rather somewhat removed without the skirt of woods lining the river bank. During our absence from the main body a strong war party of Gens des Feuilles had conducted themselves in a very insolent manner towards our fellow travellers, demanding tobacco & whiskey, ordering them to go another way that the buffalo might not be driven off &c. After amusing them more than an hour, hoping myself & companions would come up, the Major left them. The Indians, awed by his firmness, committed no violence, though at first they discovered disposition to oppose & one of them cocked his gun.

We made our supper of the spoils of the hunt, which we had brought fastened to our saddles. We found the elk-meat, though inferior to that of the buffalo, a lighter colored & better kind of venison; the chief objection to it was the too great taste of tallow. The udder was delicious, perhaps the more because our meal had been postponed & the appetite consequently increased. It was cooked in our usual way by inclining to the fire the sharpened stick thrust through it & stuck into the ground. Meals that I have ate with most zest have been those prepared in this rude manner, no doubt borrowing much from attendant circumstances. Cookery as an art is built up of the inventions of Civilization aiding languid appetite or ministering to the depraved palates of epicures. Among Savage nations all culinary improvements doubtless proceed from necessity in the first instance: to render less noxious or less unpleasant some kinds of food scarcity has reduced them to, or as a mean[s] to increase the quantity. Clear day: forenoon wind NW, afternoon SWd. Air at sunrise 62°, at Meridian 75°. Distance 22 miles.

The Journal of Stephen H. Long, July 29, 1823

Tuesday, July 29,1823. Apprehending no friendly disposition of the indians towards us from the reception they gave us as well as from their known character, we took every precaution in our power to protect our horses and baggage from depredation. The former were pinioned with long lines, or hobbled, and then arranged in such a manner as to answer the purpose of a breast work. A guard consisting of all our military strength, viz. 13 soldiers of the army was posted, forming two reliefs of six men each under a sergeant. Most of the gentlemen of the party kept watch, and so much caution was observed by all that a lurking enemy could have found no encouragement to annoy us. Still feeling that our position was insecure and that some difficulty might accrue for the want of an interpreter well skilled in the Sioux Language, and still more from the want of presents to satisfy their exorbitant demands, it was deemed most prudent to avail ourselves of the light of the moon, which arose about 10 p.m., and resume our journey by night. We accordingly started at midnight with as little bustle as possible, and indeed were so silent about it that one of the indians who slept in camp with us was not awakened till we were about departing.

Travelled 12 miles over an unvaried plain and encamped at sunrise for breakfast on the bank of the river. Proceeded 8 miles farther and dined just below the entrance of the Fol[le avoine] or Wild rice river, which rises near the Coteau de Prairie at the distance of about 40 miles NW of Lake Traverse. Encamped a little before sun-set on the bank of the river. Dist. 33.

The Journal of James E. Colhoun, July 29, 1823

Tuesday, July 29, 1823. Apprehensive from the behaviour of the war party that they would steal our horses if not attack us before day, we left the camp at midnight & proceeded on our tour. After travelling 12 miles we stopped to breakfast on the river bank. At 8 miles beyond, the Pse [*Wild Rice*] River comes in from the other [*west*] side. Where we encamped the Red River is 20 yards wide & has a current of about ½ knot an hour: the banks very boggy & the water thickened with particles of the rich light-blue clay through which is flows. No valley whatever to the river: on its banks, a flourishing growth of oak, elm, cottonwood, aspen &c. but no sugar-maple, walnut, hickory or beech. Travelled through flat rich prairie from one bend of the [Red] River to another. No granitic masses since the first day we left Lake Travers.

I repeatedly observed buffalo wallowing, throwing up the dust so as at a great distance to resemble the spouting of a whale. I killed a bull today that on counting the rugae on the horns proved to be 26 yrs. old; 4 rugae were allowed for the first year: it is probable that the domestic cattle never attain so great age. It is our practice to drive the bulls ahead of the party & kill them where we are to pass, so that there is little detention, especially as we take only the tongue & liver, rarely the marrow bones. We have not been able to procure a single cow, though in the forenoon I chased one in a band till my horse was tired & all my balls expended without wounding her sufficiently to disable her.

In this country a horse is valued in reference to his capacity for "running buffalo." If pushed, a buffalo soon tires down, otherwise he will run a long time. The hunter keeps within a few paces on the starboard quarter of the animal & aims to pass his ball from about the middle of the side to the heart or lungs. Should the discharge be ineffectual he loads while running: the powder is poured between the thumb & forefinger into the hand, clenched to form a cavity, the hand is emptied, resting the little finger & the lower part of the hand upon the muzzle of the gun: the ball is dropped from the mouth: no wadding, in general, & that is the reason the buffalo appears to be so hard to kill. In this kind of hunt, it is evident that the rifle cannot be used to advantage & therefore it is a species of firearm not to be seen in this country. The danger in running buffalo in herd is that by advancing too much on flank, a part may turn short & run over the pursuer, & if one were to fall from his horse after having wounded a buffalo, the enraged animal would be apt to gore. But I must think that the difficulty of killing the buffalo & the danger to be apprehended from him are greatly exaggerated. It is true that in one place, at least, he appears to be invulnerable: Lt. Scott fired a musket at the head of one within ten paces, but the ball did not penetrate. Buffalo have [a] larger frame than domestic cattle & though uncouth in the fore parts, they are handsomely formed in the hinder.

At Lac Travers it was estimated that the cows generally furnish from 250 to 300 wt. of meat, not counting the head & some other parts. Of the flesh, that upon the bones forming the hump is most esteemed. I thought the buffalo meat superior to any beef I have ever eaten, excelling even that of the wild cattle of the Pampas of Buenos Ayres; it has not the least taste of tallow. Salted tongues are sent in considerable quantities to New Orleans & are excellent. Eight bones are enumerated as marrow bones. The flesh is chopped off them & they are thrown on the fire: the marrow, roasted in that manner, resembles in color & consistence a custard & is even richer. It is eaten out of the bone with a stick splintered at the end & requires no accompaniament. I am unable to conjecture how much the bones yield singly or collectively, but the marrow of one thighbone affords a sufficient meal to a hearty man.

The Wonderful Ride

Being the true journal of Mr. George T. Loher who in 1895 cycled from coast to coast on his Yellow Fellow Wheel

With commentary by his granddaughter, Ellen Smith (in italics)

As I Remember Grandfather...

It's been more than thirty years since Grandfather Loher died, but only in the last two have I come to know him. Even now, I wonder if I really do.

Oh, I saw him often enough as a child—he lived less that an hour away. But Grandpa wasn't an easy man to know, especially for a little girl. I was only eleven when he died, and only three when he lost the one great joy of his life, my grandmother.

Grandfather was a quiet, even a gentle person, but one to be reckoned with. A wink, or more often a frown, said enough. Remembering him now, I still feel a certain awe.

Actually, he wasn't a big man—only five foot seven—but he had a broad frame that was strong even into his seventies. His hands, large and callused, never stopped working, if only to twiddle his thumbs or twirl the ends of his bushy gray mustache. His dress always seemed the same—a black suit, a black fedora, and a gold watch chain draped across his vest. I liked to watch its links sparkle in the sunlight.

Perhaps Grandpa really wants to be friendly, I would tell myself, and just can't let go. He was such a proud man, stubborn and not very diplomatic, scrupulously honest with a puritan heritage that let him find satisfaction in hard work. But there was no fun, no imagination, no adventure in him—or so I thought. I know now how wrong I was!

Grandfather had been gone for a quarter of a century, and I had grown up, married, and had five children of my own by that day in 1969 when I learned of the wonderful ride.

I was looking through a box of mother's old photographs and came upon a yellowed clipping from the long-defunct San Francisco Call, *datelined December 15, 1895. It carried a sketch of a young man with a handlebar mustache standing beside a bicycle, and next to it was a news story headlined: "THE ACME CYCLIST BACK. George Loher, Who Went From*

Oakland to New York on His Wheel." Grandfather?

Yes, Mother said, it was he. In fact, he had written an account of the trip shortly after he got back—Uncle George still had it. She seemed surprised that somehow, in all the years I was growing up, no one had ever bothered to mention Grandfather's amazing trip. But then, Grandfather himself never mentioned it. It was just something he had done, long ago; it was over, so why talk about it?

Then she sat down and told me the story.

Grandfather was a young man at the time, twenty-nine, single, and working at Fred Becker's Central Market in Oakland. The young butchers there had some merry times together, and apparently Grandfather was one of the leading cutups. At least, on the occasion of one evening's picnic at Trestle Glen (then a popular spot east of Lake Merritt), he is said to have clambered up onto a parked locomotive, put a match to its lantern for light, then proceeded to dance several young ladies around the grassy floor. The event was commemorated in a long doggerel epic that I found among the photographs.

Cycling was the craze of the day, and some of the boys at Becker's became enthusiastic wheelers. They joined the Acme Cycling Club and enjoyed outings and competition with other local groups. But there was a big world out there beyond the Bay Area, and Grandfather and Thomas Cornell, another butcher at Becker's decided that they should see it—on wheels.

They would pedal from Oakland to Portland, then across the continent to New York. Grandfather insisted that they were not out to set any records, so one can only wonder about the unusual route. Most cross-country cyclists rode directly between San Francisco and New York, a much shorter and less rigorous trip. They also chose to travel without maps—a decision Grandfather would soon regret.

On August 11, 1895, the two young butchers rode out of Oakland on their grand excursion, accompanied to the outskirts by a farewell contingent of "Acme boys." Cornell was destined to return in a few days, but for George T. Loher it was the start of a grueling eighty-day adventure that would end at the foot of Manhattan Island.

Grandfather must have kept a diary faithfully, for the name of virtually every town, river, hotel, and person he encountered found a place in his journal. Sad to say, those notes and the longhand draft he wrote promptly upon his return to Oakland have all vanished. But a typed copy has remained in the family and forms the basis for this book.

So far as I know, no part of the journal has ever seen print. My mother recalls that many years later, in 1928, two men came to his shop one day to talk to Grandfather about publishing it, but he flatly refused. In fact, he was downright irritated that they should even suggest such a thing. Why, I don't know. His reaction is hard to understand, but then so was Grandfather.

Naturally, the journal has a lot of personal meaning for me. But from the first time I read it, I've felt that people outside the family might enjoy it, too, and I've dreamed of seeing it published. Slowly, I began to track down missing bits of information and answers to questions the journal raised. Then, in the spring of 1976, I met Patricia Kollings of Westword Associates, who shared my enthusiasm for the project. With her editorial guidance, my efforts gained momentum, and together we have developed this book. So far, everyone involved with the project has found great delight in it. We hope the reader will, too.

We have altered Grandfather's original text only in minor ways—omitting lists of small settlements passed through and a few routine remarks about meals or road conditions, rephrasing an occasional passage for the sake of clarity, and correcting misspellings where possible. In case of doubt, however, we've let Grandfather have his way with the spelling, no matter how questionable.

We've resisted the temptation to delete or soften any of his comments out of fear that they might sound prejudiced. After all, Grandfather was a man who had opinions on everything and expressed them frankly. We hope the reader will remember that these are statements from another era, made by a young man suddenly confronted with a wide assortment of people and cultures that were strange to him. Grandfather himself might see things quite differently now.

Anyone undertaking an adventure like this today would no doubt record his experiences on film, but the cameras of 1895 were hardly pocket models, and a cyclist had to keep weight to a minimum. Nor was Mother's box of photographs much help; It didn't contain a single snapshot of Grandfather during those years. There was the wedding portrait from the following year, and another taken several years earlier—nothing more. Eventually, I happened upon the crucial picture of Grandfather and his wheel in an 1895 issue of the old Bearings magazine, and Louis Stein of Oakland, an authority of transportation history, located a photo of the Acme boys in his own collection.

Lacking original photographs of the trip itself, I have tried to reconstruct the America Grandfather saw with appropriate pictures from historical societies, libraries, newspapers, and individuals along his route. In the process I have also learned a lot of interesting, often amusing sidelights that have made the journal even more enjoyable for me. To share these with the reader, I have added personal comments here and there, printed in the same italic type as this introduction.

As the journal goes to press, I'd like to say one last, private word. Thanks, Grandfather, for taking us on this wonderful ride through an America we'll never know. And thanks for letting me get to know you through your journal as I couldn't face to face. I hope you won't be upset by our publishing it now that you've gone. We just want the world to know what a determined butcher boy from Oakland could do when he set his mind to it.

Ellen Smith
April 1978

Friday, Sept. 27

Sentinel Butte was the next stopping place after leaving Wibaux, and my ride between the two places was comparatively easy. Sentinel Butte is a section house and derives its name from a large hill that towers way above the rest and stands directly at the back of the house. It can be seen for miles around.

From Sentinel Butte to Andrews, another section house at a distance sixteen miles, the roads are not so good, as the country is exceedingly hilly. The wind was blowing a gale, and I could hardly stand on my feet. From the top of the many high points I reached, I could look down into the depth below and gaze upon a sight that was bewildering as well as wonderful. I wondered what I was going to see next.

I reached Andrews a little before sundown. The accommodations were excellent, and the people endeavored to make my one night's stay as agreeable as possible. Indeed they seemed to be quite in love with my company. The lady had to make a bed for me on the floor, as travelers did not often stop there overnight.

The wind blew a perfect hurricane that night and will leave an everlasting impression on my memory of my one night's stay in the heart of the Badlands.

Saturday, Sept. 28

The folks there expected snow, and that set me thinking as to whether I would have to prolong my stay or not. Fortunately, the next morning dawned bright and clear, and I was on my way to Dickinson before seven o'clock.

The next place after leaving Andrews is called Medora. It is situated on the Little Missouri River. Medora is the place where the Marquis de Mores, a French nobleman, undertook to show western cattlemen a thing or two by sinking two million dollars in a vain attempt at carrying on an extensive packing establishment in the heart of the Badlands.

I had a pleasant chat with the telegraph operator, who told me that Medora had been a lively place until the Marquis finally got onto himself and quit.

The operator seemed tired of his surroundings. I could hardly blame him, as Medora is one of the most dismal-looking places on the face of the globe. He asked me for information in regard to California fruit farms, as he intended to go there and invest in the near future. I gave him all the information I was capable of and resumed my journey toward the rising sun, following along the track to within four miles of Sully Springs, where I came to the road again.

I rode out of the worst of the Badlands and onto a hilly prairie at Sully Springs, continuing over the same kind of ground to Belfield, where I had dinner at the section house.

I had been sleeping and eating at section houses for two days now and had become somewhat used to them. The Belfield house was occupied by an Irish family. The nationalities differ through the Badlands. This Irishman was possessed of some half-dozen children, and on my entering the house, they amused themselves by throwing my wheel down and putting the finishing touches on my already crippled handlebars. This necessitated my hunting up another piece of iron and some more wire in order to get them in condition to go the remaining twenty miles to Dickinson. I finally managed to get them wired up so they would hold and reached Dickinson at three o'clock, with five spokes out of the rear wheel and a broken handlebar.

The local telegraph operator found bicycle repairing quite profitable during his leisure hours, and it was to him that I was compelled to apply in order to have my rear wheel steadied up a bit. The rest of the afternoon was passed by the operator and me in accomplishing the most difficult undertaking of sticking tandem spokes into ordinary nipples, his supply of spokes being somewhat limited. Repairing the handles was out of the question—that was entirely out of his line of work. I was thus compelled to leave the following morning (Sunday, September 29) with the handles no better than I had fixed them at Belfield.

Sunday, Sept. 29

While at Dickinson, I stopped at the Villar House, a so-called first-class hotel, but I found it nothing more than an ill-reputed gambling den, where one could be accommodated with any game of chance.

North Dakota is supposed to be a prohibition state, but one can have all he desires in the way of intoxicating liquor. It is served from bars in the same manner as elsewhere. The state derives no revenue whatever, and it appears very much as if the saloon keepers were at the bottom of the prohibition.

The country in the vicinity of Dickinson is of a decidedly rolling nature, but not so much so that it prevented my making pretty good time. The road was first-class as far as Glen Ullen, a small town fifty miles east of Dickinson, at which I arrived in time for dinner. A Catholic church located here is the means of bringing farmers for miles around on Sunday to attend service. Some of them require two wagons to carry their families, owing to the yearly increase.

I rode pretty close to the railroad between Dickinson and Glen Ullen, but after leaving Glen Ullen, I left the track and entered into the hills. Now I saw an almost unproductive country; when I say almost, I mean that they raise a crop about once in seven years.

This part of North Dakota is mostly settled by Germans, whom the railroad company enticed out there by sending agents to Europe to offer every inducement to get them settled on its lands. After that, the Germans are left to eke out a miserable existence the best way they can. It is almost impossible for them to make both ends meet, owing to the severe winters that prevail in this part of North Dakota. They live in dirty, low adobe sod houses, which reminded me very much of the old missions in Southern California. I lost my way several times that afternoon and had great difficulty in finding it, on account of the settlers' inability to speak English.

Not only the railroads but the state, individual towns, and private land companies all promoted the glories of North Dakota to German immigrants. In fact, the capital was named Bismarck in the hope of attracting German investments.

Monday, Sept. 30

I reached New Salem that evening at five o'clock. It was long before sundown, and I could have kept on, but as there is no town within twenty-eight miles of there, I concluded to stop for the night. I did not wish to be caught on the North Dakota prairie among foreigners after dark.

I was rewarded for stopping by meeting a young fellow from California. We spent a most pleasant evening talking over the advantages of the Golden State, and we both wondered that people live in such a barren state as North Dakota.

The next morning saw me on my way to Mandan, twenty-eight miles distant, before seven o'clock. By this time, my wheel was commencing to show the excessive riding it was subjected to; consequently, I reached Mandan with the front part broken in two. Unfortunately, Mandan was not possessed of a repair shop, which necessitated my going to a tinner, who plastered a piece of sheet brass over the break. When finished the front of my wheel looked as if someone had shot a ball of hot lead at it. It was the best I could do, however, and after eating a somewhat hurried meal, I proceeded in the direction of Bismarck.

After trundling my wheel over railroad ties for some three miles, I reached the railroad bridge—the only means of crossing the Missouri River at that point—only to be informed that I would not be allowed to cross. I was compelled to return to Mandan and procure a permit from the division superintendent, after which I crossed without further interruption.

It was three o'clock when I left Mandan the last time, and on arriving at Bismarck I discovered it was five o'clock. The distance between the two places is only five miles, and I could not imagine how I had come to consume so much time, but owing to the excitement of having my wheel repaired and the inconvenience of crossing the river, I had forgotten that the time changed—I had lost an hour between the two places.

Bismarck is the capital of North Dakota, and the capitol building is not much larger than a wealthy man's residence. There is an amount of rivalry between Mandan and Bismarck, as one can easily see when there is not even a wagon bridge across the river. The two cities each have a population of two thousand.

My stay in Bismarck was of short duration, as I failed to see anything there that attracted my attention. So I proceeded eastward, passing the state penitentiary—another insignificant-looking building—a mile from town. I reached Mendotan, the first section house east of Bismarck, after passing through some very good farming land, and stopped there for the night.

Tuesday, Oct. 1

It was the first day of October when I bid Mendotan good-bye. Proceeding over excellent roads through Sterling, Steele, Crystal Springs, and Dawson, I reached Tappen, a station on the railroad put there for the accommodation of a large cheese ranch. None of the places that I passed through had a population to exceed two hundred.

The folks I had dinner with at Tappen were exceedingly unsociable. I asked question after question as to the country's resources, and so on, but I was answered invariably in a snappish manner. I cannot account to this day for the unsociability of the North Dakotans. Perhaps they were averse to bicyclists, or upon second thought their surroundings might have had something to do with their lack of hospitality. Indeed, I cannot blame them—just think of sowing crop after crop, only to have them killed by drought or burned up by prairie fires.

A great many of the towns I have mentioned are boomtowns and are scarcely inhabited now. Poor people went there and spent what little they had in a vain attempt to raise a crop. Now they are compelled to stay and do the best they can, while others more fortunate have picked up and moved to more congenial localities.

The roads were miserable after leaving Tappen, being somewhat sandy and rutty. After I had ridden some ten miles or more I discovered that my forks were breaking again.

Jamestown, fifty miles from there, was the only place I stood any chance of having them repaired. I happened to have a few spools of copper wire in my bag, which I put into very good use, and succeeded in temporarily mending them.

I managed to reach Windsor, another boomtown, seventeen miles west of Jamestown, a little before sundown and stopped for the night. An English syndicate attempted to make a city of Windsor and built a large hotel by way of an enticer, "a la Southern California style." Their attempt was fruitless, however, and the hotel, with accommodations for three hundred guests, is now occupied by a man and his wife who have a so-called farm in the vicinity. A store and the station go to make up the rest of the once-boomed city. I had the pleasure that night of occupying one of the rooms in the Englishman's wild investment.

The hotel no doubt did look large, sitting as it did on the prairie with nothing around it but one- and two-room buildings; but W. M. Leingang, photo curator of the State Historical Society of North Dakota, advises me that it actually had only thirty-five or forty rooms—still pretty imposing for a community like Windsor, North Dakota.

Wednesday, Oct. 2

The following morning I was up and away before five o'clock, as I wanted to reach Jamestown and have my wheel repaired so I could leave there at noon. It was a bitter morning, and I suffered intensely from the cold. This, combined with the darkness that made it a near impossibility to find my way, made my situation exceedingly distressing.

I reached Jamestown a little before seven o'clock and had to wait some time for the repairer's shop to open, after which I was informed of his inability to repair the wheel owing to the absence of a brazing machine. He advised me to send for new forks, but I hated the idea of waiting in that lonely, forsaken country for the parts so told him to go ahead and do the best he could.

After pondering over it for some time, he concluded to make an attempt and succeeded somewhat in accomplishing a most unsightly piece of repairing. I also had him give the wired-up handlebars a good application of solder by way of a preventive. After having eight spokes put in the rear wheel, I was ready to continue my journey once more.

True, my wheel was a sight, but appearances counted for little with me at that time, and I was beginning to have grave fears that the wheel would not carry me through. At any rate, I intended to ride until it would carry me no longer, when I would have given it up for a bad job and have charged it to the wheel.

After partaking of a hurried dinner, I bid Jamestown good-bye at half past two o'clock and proceeded against a strong southeast wind. The wind had the effect of retarding my progress to some extent, but notwithstanding I managed to reach Sanborn, a little prairie town twenty-six miles east of Jamestown, a little before sundown. Unfortunately, I stopped at one of the worst hotels I had stopped at on the trip. I could scarcely get enough to eat, and that, combined with a carpetless floor, a hard bed, and not enough blankets to keep me warm, left a very unpleasant remembrance of Sanborn in my mind. The proprietor had the audacity to charge me at the rate of two dollars a day for the accommodations.

Thursday, Oct. 3

I was up the following morning at five o'clock and intended to ride to the next place for breakfast. But owing to the light supper the previous night, I came to the conclusion that I was too hungry to proceed and waited until six o'clock for breakfast. This meal was far worse than the previous one; indeed, I had to run the feet very nearly off the poor, unfortunate waiters before I could get half enough to eat. It was nearly seven o'clock when I had paid my bill and was ready to continue my journey against another southeast wind. I also managed to lose my way, which compelled me to cut across fields and climb fences before reaching Valley City, a town somewhat larger than Sanborn.

On arriving at Valley City, I stopped for a drink, and in doing so I was accosted by some North Dakota farmers. They asked me what I was riding for. When I told them I was out for health, pleasure, and adventure, they did not seem to see any sense in riding across the country on a wheel for any such purpose and declared they would rather walk. I came to that conclusion also, as I would not see how they could invest in a wheel when they only raised a crop once in seven years, and nine times out of ten the chances would be of that being swept off by a prairie fire before it could be harvested.

Valley City is nearly as large as Jamestown. It is prettily situated in a canyon and cannot be seen until you reach the brow of the hill before descending into the town. It has a population of fifteen hundred, and the town is supplied with water from the Sheyenne River, which flows through it. The state normal school and a branch of Keeley's Institute for the cure of drunkenness are located there.

After a hot argument with some Keeley graduates on the liquor question, I left Valley City and proceeded up the Canyon for one-half mile, after which I turned left and rode under the Soo Line [*Minneapolis, St. Paul & Sault Ste. Marie Railway*] trestle. I managed to lose my way again and did not know it until I had come to an obstruction—a house occupied by an old Irish woman. She directed me to the right road, but I succeeded in getting lost again before noon.

The country is somewhat level after leaving Valley City, and one can see the settlements or towns along the railroad for miles. But the wagon road diverges from the track at intervals and connects with forks or branch roads that run in every conceivable direction. This was the means of bewildering me to a great extent, hence the pleasure of losing my way a number of times before reaching Buffalo, a town of five hundred inhabitants, in time for dinner.

I had dinner at one of the so-called poolrooms and restaurants combined. They are nothing less than the lowest of rum shops, with the sign "Poolroom and Restaurant" over the door for a blind, in order to evade the law. I can say without exaggeration, that North Dakota contains more saloons in proportion to its population than any state in the Union, and still is called a prohibition state.

As he crossed North Dakota, Grandfather seemed to be equally critical of the illegal saloons and the temperance fanatics. Probably he was disturbed more by hypocrisy than alcohol, because he was not a teetotaler himself, although in the journal he admits to drinking whiskey only for medical purposes. My uncle Frank recalls that he kept a bottle of Old Crow on hand (the only liquor in the house) to be trotted out when business associates would call. He always drank it straight and never had more than one drink.

The wind had gained in velocity while I was eating dinner, and as I emerged from the saloon, preparatory to taking my thirty-six-mile ride to Fargo, it was blowing almost a gale.

The country is as level as a floor between Buffalo and Fargo, and the roads are excellent. Wheat is raised in large quantities, and the farmers could be seen for miles around plowing for the spring sowing. I was commencing to ride through the farming portion of the state at this time, which had the effect of making my surroundings somewhat pleasanter. It is impossible to raise a tree on those prairies owing to the severe winters, and one does not have the pleasure of gazing on any foliage until Fargo is reached. Even there trees are what might be called a scarcity. How different from beautiful California, where the farmer is not compelled to plow in the fall in order to get his crop sown as soon as the snow is gone in the spring, and where the beautiful foliage can be seen on all sides, growing to dizzy heights and proportions! How I longed to get only one glimpse of California's trees as I rode over these treeless prairies!

I reached Fargo that evening a little before six o'clock, after passing through Casselton and the famous Dairymaple ranch. This ranch consists of twenty-five thousand acres under cultivation. The soil is a rich, black loam, and wheat and flax are raised in large quantities.

The roads were excellent, both after leaving Casselton and while passing through the ranch but I was completely exhausted from the continuous strain I was subjected to while proceeding against the wind.

The North Dakota race was being held at Fargo on my arrival, and the city was crowded with visitors. I had a little difficulty in finding accommodation but succeeded at the New Fargo Hotel. I had ridden seventy-five miles that day, and I was commencing to go through better country and over better roads. This gave me some encouragement, notwithstanding the battered condition of my wheel.

Fargo is the metropolis of North Dakota and claims a population of ten thousand. It is situated on the famous Red River of the North, which divides North Dakota from Minnesota. The authorities are very strict with the prohibition law in Fargo and will allow no saloons to run openly, as is the case in the smaller places through which I passed. One can get all he wants, however, if he knows the ropes; and if he is a stranger, all he has to do is to walk over the Red River bridge to Moorhead, Minnesota, where he can quench his thirst in a jiffy.

I had followed the Northern Pacific Railroad from Spokane Falls to Fargo, a distance of 1,261 miles. I was advised to leave it at Fargo and follow the Great Northern, as by so doing I would avoid the sand through which the Northern Pacific runs east of Fargo.

Friday, Oct. 4

I left Fargo and North Dakota on the morning of October 4. It was but a few moments after leaving that I crossed over the Red River bridge and into Minnesota.

After passing through Moorhead, I turned to my right and rode over exceedingly level country in a south-easterly direction. Sabin was the first town I reached on the Great Northern. I only stopped there for a drink of water and, proceeding on, reached Barnesville, where I commenced to ride over rolling country once more.

At Barnesville I inquired for the road and was advised to follow the track, when I should have kept away from it for some three miles. I did not discover the mistake until I had run into a swamp where there were no roads whatever. I endeavored to ride after reaching the swamp, and in doing so ran into a large hole, breaking five more spokes out of the rear wheel. This was a sufficient cause for making me walk the remaining three miles to Lawndale, the next station, where I was directed to the right road.

Lawndale consists of a grain elevator and a store. I was informed that the occupants of the store fed travelers, should they happen along at mealtimes, and as it was noon, I applied for dinner. The cook happened to be indisposed that day and politely offered me some cheese and crackers. I did not propose to wrestle with a bicycle over bad roads on a diet of cheese and crackers, so I proceeded to Rothsay, twelve miles distant, and had an excellent dinner at the Rothsay Hotel.

When I inquired for the best road to Fergus Falls, the next important town, the hotel proprietor willingly informed me that I would have to go six miles south then turn and go seven miles south again. I looked at him in amazement but did not dare to say anything for fear that I might be laughed at for my ignorance, not knowing but what that was the Minnesota style of directing a stranger over strange roads. I mounted my wheel and proceeded in a southerly direction with the intention of solving the road question myself.

I was in the farming country now in every sense of the word, with vegetation on all sides of me. Consequently, there were numerous roads running in every conceivable direction. This part of the state is settled mostly by Swedes and Norwegians, which made it difficult to find my way, owing to their inability to speak the English language. This was indeed a most serious predicament, as I discovered after leaving Rothsay.

After I had pondered my situation for some time, a German farmer, who had mastered the English language to some extent, came along and advised me to keep straight ahead until I saw the smoke from the asylum at Fergus Falls. I took his advice and proceeded for some distance but saw no sign of smoke.

I was beginning to lament my ill luck for not being supplied with a field glass, when I discovered a dark line of smoke rising heavenward to my left. How to get over there without taking the wrong road was another question for me to solve. The first turn I made brought me up to a farmhouse, where I found that the road ended. This necessitated my return to the main road for another attempt.

I had the unsought-for pleasure of repeating this a number of times before reaching Fergus Falls, at which place I arrived at five o'clock, after having to wade through a number of creeks on the way. Just as I was riding into town, a Swede commenced to swear at me in his native tongue for scaring his team. I failed to understand what he said, and perhaps if I had, there would have been a row, as I was not in the best of humor at the time.

Wonderful Ride Maps

George T. Loher's Wonderful Ride Through North Dakota

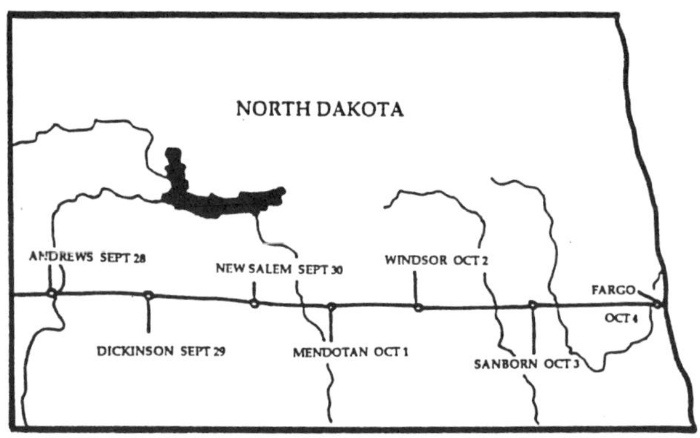

The Wonderful Ride of George T. Loher
August 11 – October 30, 1895
San Francisco to New York City

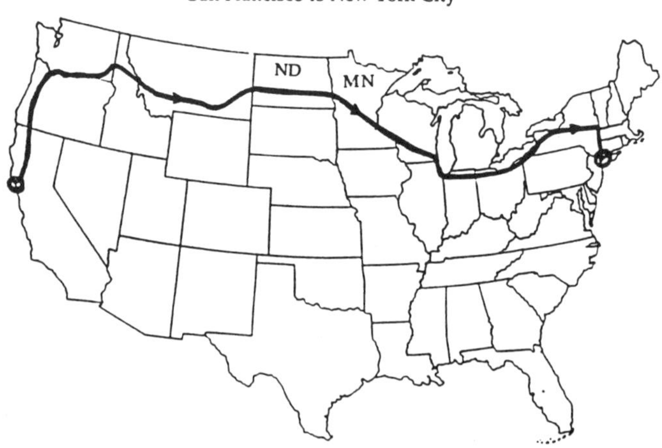

Homestead Days

Sods, Logs, and Tar-Paper: Memories of North Dakota Homemakers *is the first in a series of three books growing out of an oral history project. The project, "Memories of North Dakota Homemakers," was undertaken in 1989 as a part of the celebration of North Dakota's state centennial. The excerpt printed here is from the first chapter of the book.*

My parents were pioneers in the territory. My father came from Germany while my mother was born in Wisconsin.

Were they homesteaders?

Oh yes, my mother was a homesteader as well as my father. While they were living on their land, that's how they got together.

Esther Schoewe, 85, Barnes County

THE HOMESTEADER

The Homesteader's day was more work than play
And the joys that he had were little
But when he came to the end of the day
He was strong and fit as a fiddle.

The Homesteader built his house from sod
The same as the barnyard swallow.
Then a church to worship almighty God
And a schoolhouse were soon to follow.

His life was expressed, not in words, but deeds
And calls from his neighbors were heeded.
He broke the soil and planted his seeds
For the strength from the soil he needed.

He reared his children and built a state.
His rewards were not all in money.
There were clouds at times but a kindly fate
Brought days at the end that were sunny.

So hail to the Homesteader — man and wife
Dressed in calico clothes of cotton.
They lived a busy and hard-working life
Which ought never to be forgotten.

As long as men live, they will always know
That the sunset skies will be redder
And all of us bask in the afterglow
Because there was once a Homesteader.
Clell G. Gannon

I was born of mixed heritage I guess. My mother was Norwegian and Swedish, my father was English, Irish, Scotch, and German. So I always told my children that they were a duke's mixture. My father was born in a log house in Minnesota at Faribault. My mother was born at Tilby, not too far away. I'm sure that they didn't know one another until they homesteaded in Burke County. My dad homesteaded in 1903, but I'm not sure when Mother homesteaded. She had a quarter of land not too far from where my father's quarter was. And she worked this land herself to prove up on it, as did my father. Then they were married in 1907.

You mentioned that your grandparents lived a short distance from you. Were they homesteaders too?

Oh yes! In 1903 they homesteaded at the same time. And my aunt tells the story — because the grandparents had a large home, anybody that came from Minnesota always stayed at her parents farm while they were getting a shack or a house built for themselves. Many of the people that live back in the hills from us, stayed at their place while they were getting their places ready. We can still see that trail through the coulee that was the road up from the flats into the hills where those homesteaders lived.

Do you have any recollections that your mother told you about how they immigrated to this country?

Well no, not my mother. But my father tells about how they come with a box car on the railroad. They had all the farm animals that they were bringing with them, besides the lumber for this big house that still stands on that place. It will be 83 years old this year and it's still being used.

Ruby Bryan, 73, Burke County

My mother's parents had come from England into Canada and were living in Quebec where my mother was born in 1883. When she was three years old, her parents decided to come to North Dakota. They rented a box car and loaded their household possessions and a few cows and came by railroad to Devils Lake, Dakota Territory. This was the end of the railroad line at that time. After arriving at Devils Lake they made arrangements to drive a team and wagon to the area near Towner. This was in 1886 and Towner was just starting. Grandfather left the family at Pleasant Lake, there was a sort of way-side home there where the family could stay while he went north of Towner where he was looking for land to settle on. He followed the Mouse River for about twenty miles north of Towner 'til he found some land he thought would be good. This was what was called a relinquishment, some one started to prove it up, then gave it up. He bought it, then went on and proved it up. He brought his family, which at that time consisted of four children. They made their home then in that area and the family grew and prospered. I am not sure Grandmother was always happy, she had children to care for and she was very frightened at times because the Indians roamed the area and they came to the house and frightened her so much.

When Mother was twenty-one years old she proved up a quarter of land that joined the one that her father owned at that time. She lived in a claim shack to prove it up. Things that frightened her while living in the claim shack were bad storms, these would come up at night. She would try to block out the lightening by putting blankets over the windows and so on. On December 30, 1909, she received title to the land and on the same day she deeded it to her father. That gave him two quarters of land. She was going to be married, and in 1910 she married Harold McCombs. I am not sure why she wanted to prove up this land, unless it was to help her father to have more land.

Vera Fairbrother, 71, McHenry County

My grandfather, Andrew Hill, came from Ontario and homesteaded in 1889. They lived in a log house 'til they were able to build the frame house on the hill, where we all lived a long time.

The town of Hannah was eight miles east of the farm. They homesteaded on about 80 acres. They were both Irish and Scotch, I guess. My mother used to say, "We're Irish and Dutch and don't amount to much".

Donalda Wheeler, 71, Cavalier County

My folks came from the Province of Quebec in Canada.

Your mother and dad were both full blooded French, then?

Yes, all French. He settled in the Belcourt area, in Shell Valley Township. I think he just had a small log cabin at first.

Irene Leonard, 78, Rolette County

My grandfather, John McBride, was born in Ontario in 1850. In 1881 he came to the Dakota Territory and homesteaded in East Alma Township. His first home was a pole construction, banked with dirt and it had a sod roof. He established a post office in his home known as Alma Post Office. Sometimes he walked to St. Thomas to get his mail. His wife, Elizabeth Mickle, and four children came to Dakota in November of 1882. They came to Pembina first, then to Cavalier. John met them there and brought them home on a stone boat pulled by oxen. Grandfather loved politics. When 75 North Dakota citizens met at Bismarck on July 4,1889 to draft the Constitution of the new state, Grandpa was one of the delegates representing Cavalier county. He was elected the first Senator from Cavalier County to attend the first Legislative Assembly of the new state of North Dakota. Our thanks to Grandfather and all the others for a great state.

Grace Johnson, 73, Cavalier County

I was born September 11, 1900 in St. Elizabeth, Manitoba, Canada. My mother was from Pembina and my father came and worked on a farm on this side. His father drowned in the Red River before my dad was married. They were married in Pembina where four children were born, and they went back to Canada where I was born. I am of French heritage and can still speak the French language.

Marie DeFoe, 88, Pembina County

They came from Isle Verte, which is south of Quebec. They traveled until they found an area that they would like to make their home. That brought them to the Mississippi River, but the trees along the Mississippi River were all oaks, and the Hamels were especially interested in maple trees, because they knew the art of making maple sugar and that was quite a source of income for those Canadians. When they came my father was only seven years old. When they reached the Mississippi, they didn't like all those oak trees, they wanted to be where the maples were, and the Indians told them if they would go towards the setting sun it would be all maple trees, so they continued towards the west. Our farm was all wooded and my father's business was mostly selling fence posts. The area was just being settled and they had to build fences. The settlers around there would come with their wagons and horses and get those fence posts.

Marie Saumur, 94, Grand Forks County

We came down to Cando, North Dakota. My dad worked on a farm there north of Devils Lake, and learned to drive horses. In the winter of 1905 we went out to McKenzie county and filed on a homestead. He and one of our neighbors worked for a rancher and they each bought a horse, so then they had two horses. Then they bought one of those walking plows with the handles and they could plow up things that way.

Margaret Brenden, 89, Mountrail County

Mother came in 1902 with her parents. It was a lovely spring day in March when they left New York. Then they came on the train to Berwick and there was a three day blizzard so they had to stay in the depot for a while before her sister's husband came to get them. They were filing on land southeast of Berwick. Mother worked for a Mrs. Schuman, who had a cafe in Rugby and that's where she met my dad. Mother filed on land in Ness Township, 40 acres, and she had to stay with that for a year.

Alice Berdahl, 78, Pierce County

My dad came out here in the fall and built a house a barn on one end and the living quarters on the other, just to get through the summer. They shipped out cattle and horses in the fall. We come by train and I can remember we come in March, on a stormy, blizzard day. A family from Burnstad lived 14 miles away and came with a team and I guess it must have been a sled, and then got all of us in that sled, the 10th of March, I'll never forget that.

Hilda Bohlin, 89, Emmons County

My grandpa (John Urban) immigrated to the United States from Czechoslovakia in 1875. My grandma (Kate Beneda) immigrated from Czechoslovakia in 1883. They were married in Vesleyville, North Dakota and homesteaded on a quarter of land in Cleveland Township. Sometime during the first year or two of their marriage Grandma observed the neighbors picking up buffalo bones. Grandma couldn't speak English so she couldn't find out why the neighbors were picking them up. She thought there must be somebody buying them when they were all picking them up. She told Grandpa about it, but he didn't believe anyone would be buying bones. Grandma took a sack and started picking bones too. When Grandpa saw her he was very upset and told her not to pick up the bones. So Grandma went picking bones on the sly. When Grandpa was in the field she would go through the coulees. She put them in a pile in the coulee in some brush. After the seeding was all done, Grandpa was going to go to Park River, about 20 miles away. As he was preparing to go with the wagon and oxen, Grandma told him she had this pile of bones in the coulee. She asked him if he would pick them up and take them to Park River. He didn't want to take them. Grandma said, "How are you going to buy groceries, you have no money".

Grandpa said he'd get a mortgage on the crop he just planted. Grandma begged him to take the bones, he finally consented very grudgingly. Grandma helped him load the bones in the wagon. She was afraid he'd throw them out before he got to Park River.

As Grandpa was approaching Park River he saw piles of bones as large as straw piles, so he knew someone must be buying them. When he arrived in Park River he inquired and found someone that was buying buffalo bones. The bones were weighed and Grandpa got $17. for them. He thought he would keep the $17. so he'd have some cash on hand. He walked into the grocery store and asked if he could get some groceries, he would give them a crop mortgage. The owner said, "No, it is too early in the season to take a crop mortgage. Grandpa had no choice but to go easy on the groceries and buy what he could for the $17. When he came home he said to Grandma, "I don't know what I would have done without the bones." Grandma said, "I could have picked a lot more if you would have let me."

The buffalo bones were shipped out on the railroad to be used for purifying sugar. This is one of the many pioneer experiences Grandpa and Grandma Urban used to tell us about. They lived on this same quarter of land until they retired and moved to Rankin in 1916 and their son took over the family farm.

Rose Potulny, Walsh County

I was born near Alden, Minnesota, not too far from Albert Lea, my folks used to rent a place there. So, when they heard they could get free land up in North Dakota — a 160 acres of it — they just got the urge to go up here. So, of course, my dad and four other men went up here in the fall of 1898 and homesteaded, they called it "filed" I guess. They had to go to Minot to take care of those papers.

Then they came back and in the spring, my dad packed an immigrant car with his horses — only two of 'em — and some feed and a little bit machinery and a little bit household goods. He shipped that up and he went with the immigrant car to take care of it, and he had a cow or two. I was just a kid, not quite two years old, I can't remember it. My brother was four and we came on the passenger train. When we came into Bowbells on a stormy day in April there was my dad to meet us. He had gone out to the homestead about twelve miles from Bowbells and was in the process, with the help of some neighbors, of building a shack to live in. He got us in the sled and drove across the field, the snow was way up over the legs of the horses some places. It was a dreary sight, I guess. I don't remember it. When they got out there, they still didn't have the windows put in this two room building but they hurried and got them in and then we were established in our new home.

Where did your parents come from? Did they immigrate to Minnesota from another country?

They both came from Denmark. My mother was from the northern part of Denmark and my dad from the southern part, but they had never met in Denmark. My dad came over here and he worked as a carpenter. He had learned to do carpentry work in Denmark. He came over to Michigan and did carpenter work there together with his brother. They built a church and a high school and different things. Now I'm telling you, they didn't have electric saws, they didn't have electric drills, they did everything the hard way, but they did good work. There is a church he helped build nearly 100 years ago and it's still standing in Big Rapids, Michigan and it's still in use. Then after a while he came to Alden, Minnesota. Mother had come there some time before and there she met my dad. After a while they were married, in 1895, and they farmed a rented place there. Of course people are not too happy about farming when they had to rent the land, so that's why they got the big urge to come to North Dakota when you could get FREE land. We were the first family that was brought up here, my mother and we two kids.

Did your mother and dad ever tell you about leaving the old country and crossing the ocean?

Yes, they crossed the ocean and it took a long time. Mother came across on a ship and when it was halfways across it lost it's anchor, or "screw" they called it, and it started taking water. They couldn't get out in life boats and get very far, of course, but they made a signal with flags to an empty cattle steamer going through there. They signaled to them that they were in distress, then they signaled back to put their lifeboats down and get as many of the women and children in the first boats and take them across to this cattle steamer. Next they took the older men, and so forth and they got all of 'em across without even one of 'em drowning. The next day they could see the ship that they were coming in was hanging at a "list", and within a few hours it went down. Mother, all she got a allowed to take on this boat — was a little hand bag with her thimble and scissors and a little money and a few handkerchiefs. That's all she brought with to this country.

Anna Sorensen, 85, Burke County

What were some of your hard times when you first came over?

When we first come to this country? Well, it was pretty hard everybody was out working, so we chipped together so the folks got enough money for a team of horses and a wagon. We kids were out in the neighborhood, we couldn't all be home, there were seven children, in that little sod house.

Jacob Dockter, 95, Sheridan County

Grandpa and Grandma Adamson were early homesteaders. They had come from Denmark and came first to Iowa where they lived for 5 years. They were of Danish descent, but were raised under the German rule because they were from the strip of country called Schleswig Holstein. In 1898, Grandpa Adamson and Mads Olsen and his son, H. C. Olsen, and Jeppe Skow came to Bowbells and filed on their homesteads in the spring. Then in the fall, I think it was the latter part of September, they moved up here. They didn't have much time to get a shack put together, and build little sod barns for the cattle and stuff that he brought with.

Where were your father's parents from?

They came to Minnesota first, then Dad, in 1904, came to North Dakota. He was a carpenter by trade so he worked in Bowbells on buildings that were put up on Main Street, and the school house.

Margaret Lien, 69, Burke County

Dad had homesteaded in 1897 on the southwest quarter. Then when they proved up, they had to go to Devils Lake with a wagon and team. He and the neighbors went together to prove up.

Fred Dubuque, 83, Rolette County

Your parents were immigrants from what country?

From Germany. My dad went to German school and they came from Germany. My mother's parents came from Westfreisen, Germany and they brought Mother along when they came. She must have been about school age. They settled in Iowa, and shortly after they arrived, Grandmother died. Some English-speaking people took my mother in and sent her to school, and she worked for her board. She went though the grades in Iowa.

Emma Edinger, 86, Wells County

My mother was born in Germany, she was 16 years old when she came across on the ocean, it took them six weeks to come across. She spent her sixteenth birthday on the ocean.

Eva Eastman, 87, Stutsman County

Tell us about when you came to America.

We came over in 1914, just before the war started. We were 7 days on the ship, the Hamburg American Line. "Imperator" was the name of the ship. There were four classes and we were on the third class. The first and second class was in the center of the ship and the third was on one end and the fourth on the other side. One time they were giving everybody bananas and we were feeding the fish! (laughter).

Margaret Manz, 88, Sheridan County

I was born September 11, 1894 in Lodema Township, St. Thomas, and have lived here all my life. My dad, Frank Lawonn, was three months old when they came from Germany. My mother was born near Pine Island near Rochester, Minnesota. I was raised by my grandmother, as my mother died when I was one month old.

Anna Briese, 92, Pembina County

My father immigrated from Germany and came across with his folks in 1870. His parents both died on the ship, which they called at that time, "ship fever". His uncle and aunt were on the ship with him at that time and took him to Iowa and raised him.

My mother was born in Illinois and then she came to Iowa and they both married when they were 17 years old. In 1906 they came to North Dakota. They were coming with the intention of homesteading but they could not get a homestead anymore at that time. They settled on a farm 9 miles northwest of Bottineau where they lived a couple of years, then they moved 12 miles straight south of Bottineau and lived in a sod house. It was near a small town named "Gardena". I attended school there for eight years.

Alma Keith, 81, Renville County

He [Dad] read in the Dakota German paper, that John Wittmayer of Goodrich had land for sale and the folks come to North Dakota in 1900. We landed at Denhoff just about a mile and a half or two southwest of Denhoff. We lived in a little shanty. We only had the cook stove, and Mother had it going during the day, There was no heater, it was all in one room, there was no upstairs, nothing! On the 15th of May we got a big snowstorm. Everybody had their cattle in the pastures, and their horses.

A man gave us a sack of potatoes, and they put it beside this little cook stove and the fire went out. Next morning the potatoes were frozen beside the stove and the cattle froze in the pasture.

Rosie Gries, 90, Sheridan County

Gustav A. Reddig, my dad, was born in Germany in 1892. At the age of 21 he and three other German families came to America by ship. It took two weeks. After the long voyage to New York, they were told to wait in barges off Ellis Island until they were inspected for diseases. North Dakota had just become a state three years before so that's where they headed. Gust was able to claim a piece of land around what is now Hamburg. It cost nothing as he was 21 and would become a citizen after five years if he put a building on it. After five years the price of the land was $1.25 per acre and later $18.00. Forty to eighty acres was enough to raise a family. Later he bought more land, and built a house. In 1899 he married Johanna Broschat who also came over from Germany in the same ship when she was 13 years old.

Lydia Helm, 84, Wells County

Glendon [husband] moved up here from Iowa. It was in 1906 when they brought him up here. His parents were of German heritage. His grandparents eloped in Germany and were coming to the United States because of the compulsory military program over in Germany. They got on a boat that they thought was headed for the United States, and in place of that, it was a freighter. They were 92 days coming across because they picked up freight at so many different countries. Glendon's mother was born in New York

City and her dad was a taxi driver in New York, a horse drawn taxi. Well, they were there for several years and then they moved to Richfield, Iowa. Glendon's grandfather was a Civil War veteran. They lived there in Iowa for several years and then they moved to North Dakota in 1906. He was four years old when he came to North Dakota.

Ruby Bryan, 73, Burke County

All my grandparents came from Iceland, my paternal grandfather was Jonas Hudkinson and he came to this country and took the "kinson" off — therefore, our name became Hall, which does not sound Icelandic. He settled on a homestead two miles south of Gardar in 1880, with Grandmother, two small boys, two cows, an old stove and some bedding. He built a log cabin partly underground. He farmed and educated himself, studying law. He settled some early day cases. He was also an insurance salesman and a poet. He wrote poems for a lot of occasions. He and Grandmother had six sons, my father being the youngest.

My father and two of his brothers built a large stone 2-½ story house in 1904, which still stands. My maternal grandfather is Benedict Johanneson and he settled near the townsite of Gardar. He built an earth dwelling in 1878 and it was fairly warm and comfortable. He built a log house for his family the following summer. Like the Hall grandparents, they too built a large 2-½ story frame house in 1904.

Asta Swenson, 66, Cavalier County

My great grandfather came from Ireland. He was to meet his brother in New York but somehow they got separated. Some way he made his way to Troy, Minnesota, there he homesteaded. They lived in several places in Minnesota, but they came to North Dakota in 1883 as laborers on the railroad. He came up and located a homestead — but they didn't move up until 1886. The family came to Devils Lake on the train and then they were three days on the road in a wagon from Devils Lake to Leonard Township. They had three oxen, three horses, two colts and five milk cows.

James Leonard, 81, Rolette County

My parents were of Irish and English descent. They told me very little about their heritage, in those days children were seen and not heard!

The climate of Ohio did not agree with my mother and the doctors recommended they move west to a healthier climate, naming North Dakota as a very good state so in 1907 they boarded a "HON YAK" [homestead] train bound for North Dakota. For three years they worked on the old Dr. Louis Goeschel Ranch north of Flasher. My mother rode horseback and herded cattle. In 1910 they decided to homestead and while waiting to file, my mother worked for Hal DuBerrier's Mom in the telephone office in

Flasher. Both families, meaning my folks and grandparents, homesteaded a mile apart five miles north of Shields. They built their own sod houses and they had a team and wagon between the two families.

In 1910, the year of Haley's Comet, both families left Dakota by covered wagon for Montana. In Grass Range, Montana, carpenters were needed so my grandparents stopped there. Grandpa Dilley practically built the entire town in a period of a very few years. Soon after, we came back to the homestead in North Dakota and the men, during the harvest, would go to the wheat fields to work, mostly to the McClusky area.

Bea Omber, 85, Grant County

My daddy was a full blooded Irishman, he came from Ireland, possessing the charming brogue and silvery tongue capable of coaxing blood from a stone. One neighbor said about my Uncle Denny, "He was a very great man, a humanitarian, the finest neighbor, and always a gentleman and last but not least, he had humor that just didn't run out."

This uncle, how did he come to North Dakota?

He enlisted in the army in Lincoln, Nebraska and they sent him to North Dakota to help keep down the Sitting Bull uprisings. The pioneer people were having trouble with the Indians, they felt the treaty said that pioneers were to stay east of the Missouri River, and the pioneers were settling west of the river, that's when Custer and Sitting Bull got into trouble. My uncle was a soldier at Fort Yates at that time.

Carrie Weinhandl, 70, Grant County

Father and Mother both came from Norway, my father came from a beautiful valley called Hallingdal, this was where rosemalling comes from. When he was 18 years old, he came to this country and settled in the Red River Valley where he went to work on one of the Bonanza farms. These farms consisted of thousands of acres, but all the work was done by men, and of course, horses. My mother was born also in Norway, but in the far north country, in the Lofoten Islands. There her father was a fisherman, he fished for cod and if any of you know what lutefisk is, this is where the cod comes from and is dried that makes the lutefisk. She also came to North Dakota when she was 18 years old. Since the immigrant Norwegian girls were much sought after as maids on these big farms, she came to this big Bonanza farm as a maid. It was there that Mother and Father met, married and settled down in the Red River Valley, where they lived to celebrate 52 years of married life. I don't believe either one ever went out of the valley, and neither one ever had the chance to go back to Norway to see all the family that they left there.

Many people have asked me why so many immigrants came from

Norway in the late 1800's and the early 1900's. Norway at that time was a very, very poor country. There really wasn't enough food for the people. Only 4 % of Norway can be cultivated, which doesn't leave much room for growing food, so they relied on the sea for what they ate. So this was one reason that my father and mother both came to the United States where they had heard everything was wealthy. One of the things he always told about was the first meal he had at his sister's home, by the way, two of his sisters had come before he did, and had settled in the valley. The first meal he ate in his sister's home, she kept plying him with food, he couldn't believe that he was free to eat all the food that she was trying to give him, because he hadn't seen that much food in a long time. It took from 4 to 6 weeks for the boat to come over, and no food was provided on the boat, they had to take it with them. He brought a little chest full of bread and dry meat and hard cookies and that had to last them sailing clear across the ocean.

They settled on a rented farm in the valley, where five daughters were born to them. I was one. Everything was done by horses, my father never lived to own a tractor and he never had any sons, so we girls had to be the boys and three of us lived to adulthood. We learned to operate all the machinery that we had on the farm and even it was our job to break the colts for farming. We had a happy, carefree childhood and I don't think I regret any part of the unhandiness that we had. I can't think that we really missed anything. We had a very happy childhood growing up on the farm.

Julia Schlaht, Stutsman County

My mother's father and mother came from Norway. My dad's did too. My grandfather had been a painter in Norway. He did a lot of interior decorating, "rosemalling". In about 1850 rosemalling went out of style and, of course, he didn't have so much work to do. In Norway there were a lot of people and they did not always have enough to eat, so he came to — I think it was Cincinnati. He painted trim on wagon boxes and wagon wheels. Then some men friends of his came from Norway, I think there were six of them. They convinced him that he should come with them to Minnesota. There was free land if they went there. So he left his work and went there. He was the only one that had any money of the group. They had all come with just enough to pay their passage over here. They bought a pair of oxen and they bought some kind of conveyance.

When Grandma came from Norway, those men were there and she had to cook for those men. She made flat bread and they had dried fish, that was their diet. When Grandma came, they had four children in Norway. The oldest died before she left there. Then they came on just a small sailing ship. It took three months to get here, I think. They stopped in Ireland to pick up another family. The children that belonged in this family were sick — they were small children. They had what they called cholera, I guess it was a dysentery. There were a lot of small children on the ship and they all got that sickness and every one of them died. I don't remember how many they said died, there were a lot of them. So when she [Grandma] came she didn't have any of her children, but then she had seven children in the family after that. They had hard times and good times, I guess.

Myrtle Bush, 82, Bottineau County

In the beginning I planned to go back to Norway after 5 years. I should stay here 5 years I thought, but then I was going to go back.

Did you think you'd be rich then?

Ya, I thought so. I thought we could pick up money here. I was working in a hotel in Ambrose and a cowboy was shot. He rode down the street and then he got shot. I heard the shots and just coming from Norway, I thought "now I got into something".

[Anna's homestead is located south of Twin Butte church about a mile and a half. She is the only one in Divide County who still has her own homestead].

Anna Strand, 106, Divide County

Our father was very active so he got jobs in the wintertime, or he went to school in the wintertime. He got a job on the Great Northern Railroad, and I think he went as far as Great Falls, Montana. He was their bookkeeper. They doubted whether he could do this work, being a green Norwegian, but he did, and it was the best report on paying the bills that they'd ever had.

Bertha Tufsrud, 87, Pierce County

She [mother] came from Norway when she was 21 years old. She stayed with her uncle at Fort Ransom. She stayed with them until she learned the English language and she worked in cafes and hotels while she was there. In 1910 she went to the western part of the state and homesteaded there.

Did she have any relatives near her when she homesteaded?

No, they were living out here alone among the hills. There was a little shack there so evidently someone had lived there and left. I remember so well there were four of us who were born on the homestead. My mother was left alone and there were seven of us. I was only ten years old when my father died, I was the second oldest. When we got to be fourteen years old we had to be on our own except for school. We had to work during the summer. My mother was getting $56. a month pension and then when each of us got to be fourteen, $8. was marked off it. I can't imagine how she got along. Of course she had cows, chickens, pigs, and garden stuff. She had a horse and buggy, you know. She didn't want to raise us in town. After my dad died she bought this farm — a quarter and an 80. She had a house and barn built. She did a lot of things a woman shouldn't have to do, but she did it. She was both father and mother to the family.

Alice Foreng, Nelson County

I was born on my mother's homestead, which she homesteaded in 1909. My father came from Trondheim, Norway with his parents when he was seven years old. They settled at Brandt, South Dakota. Later they moved to North Dakota and homesteaded in 1906, the adjoining homestead of my mother's. My dad homesteaded north of Ruso, in McHenry County. He received his patent on May 15, 1911. In 1916 he traded his homestead for a top buggy and a team of black horses. According to the map, the Sully Expedition of 1885, no doubt, watered at the spring on Sec. 36-144-73. This is in our pasture. There is also what appears to be a grave on the southeast quarter of Section 2-143-73. This is also in Atwood Township which was named for William Atwood, the first resident of the township. He filed on a homestead in 1903.

Edwin Johnson. 71, Kidder County

My dad was born in Elverum, Norway and my mother was born in Valdres, Norway. My mother came when she was eleven and my dad was sixteen. They both homesteaded northwest of Dunseith. My dad had homesteaded already and she [Mother] came up from Wisconsin and homesteaded. Her brother encouraged her to come to the Turtle Mountains and stay with him. He told her about all the land that was available, so she and a friend came up and they both homesteaded. This was near Willow Lake and my mother and dad met at a party there. In 1907, they were married at the Wenstead School west of my dad's homestead, by Pastor Svingen.

My mother's homestead was sold to a man who owned the land adjacent to hers. Her house was moved from her homestead to my dad's homestead and that's where we were all born — on my dad's homestead in that house.

Alice McKay, 78, Rolette County

My grandparents had come from Norway, all of them. My mother's parents came on a sail boat in 1850, settled in Wisconsin and then moved on to Iowa. Mother was about 20 years old and came up here and worked and married my father — they lived here since 1879.

Verna Johnson, 74, Walsh County

Henry Brekkus, [dad] came to America in 1892 from Brekkus, Norway. He came to Lake Preston, South Dakota. He worked on farms in that community for three years, then he began farming for himself. In 1901 he came to Rockford Township, about nine miles from Norma, and homesteaded there. In the fall of 1906 he went back to Norway to look for a wife. On May 30, 1907 he married Anna Skjerveggen, who was 13 years younger than he was. They were married at the church where he was baptized and confirmed. The wedding lasted three days of celebration. They boarded the ship, a small Norwegian boat at Bergen, Norway and went to Liverpool, England, which took one and a half days. They stayed there a few days and left on a ship of the Canadian Pacific Line. After 12 days, they landed at Quebec, Canada. They went by train to Winnipeg, which took four days and from there to Kenmare, in July of 1907.

Imagine them coming to a one-room shack on the prairie where there were no trees or mountains around.

Thelma Johnson, Burke County

Katherine Misialek was born in 1877 to Vincent Misialek and Augusta Vada in Jaslo, Poland. She had a neighbor, Andrew Misialek who was born in 1875. His brothers went to America near Warsaw, North Dakota to homestead. He went later and after awhile he sent for Katherine to come. She was 24 (about 1901) when she came on the ship to the United States.

Everything went fine on the ship. The tall buildings of New York were impressive. She got on a train to go to North Dakota, and she remembers how frightened she was when the black conductor on the train offered her a nice red apple. She was very hungry, but refused to take the apple. She had been warned about strangers.

Katherine Misialek, 109, Pembina County

I was born in 1890 in Russia, in Glickstahl. We came to America in 1901 on June the 6th. We go to Germany with a wagon mit [with] a team mit one, two, three families and then we go mit a ship "Kaiser Wilhelm." We were 11 days on that ship. Ohhhhhh — there was a storm for three days. Then when we come to New York there was a big building and all nice people and then they take the money from my dad and we have that money cashed. There var no tricks and we say a dollar and that was a rubel, and they take all my dad's money. I don't know how much, and we get a half. We get a rubel, 50¢. You got $1000. and then he take $500. and you get $500. — that was not nice! Well, we could do nothing, we had to do as they say.

There were nine in the family, my father's older brothers were mit us too. There were three families in one house. You cannot lay on the bed and stretch, we had to sleep crosswise on the bed. We did not have any cows, just three horses. My dad and brothers came mit the wagon and my mother and the girls went to Glen Ullin by train. My dad hauled the little furniture we had by wagon. We got one room in that house because there were three families there. I don't know how we made it but we did.

My dad got very sick and we didn't have much money. A man came and said he would give us the money and we called for a doctor. Our mother was crying and we was crying and my father prayed for a house and we did get a house. Then my mother's father came mit three other families. They all looked for land, everything was all open. People came from Russia and other lands. We took a quarter of land and we had to stay there five years. We did not have to pay taxes on it, and after five years our dad became citizen and we, all the children were under age, so we all became citizens too.

We had no cows, no milk, no butter and no jelly. For breakfast we ate tea and bread, and for dinner we ate dumplings, strudla, knephla soup or bean soup. In the evening we ate the same, bread and tea, there was no butter. When we go to the table, we don't say, "I don't want this or I don't want that", we just ate what we had.

In the spring, my uncle built his own house and in the fall my other uncle, my dad's brother, came from Russia and they moved into that house. In the spring they built their own house and after 5 years they all had their citizenship too.

Carolina Doffee, 96, Mercer County

My birthplace was in Russia, Neudorf was the name of the village. We were Germans-Russian.

How did you come over to the U.S.?

Well, we came over as far as to New York with the ship, and from New York with the train as far as Hosmer, South Dakota. But when we came to New York the folks were completely out of money. Then they had a good friend, they had to borrow money to spend for the trip from New York to Hosmer, that happened in 1899.

Where were you able to borrow money from?

From one of my dad's best friends. He told them, "Anytime you need money I'll help you out," and he did. He gave them $260.00, that's what the folks had to borrow when they got out of the ship.

And you traveled by train from New York to South Dakota?

Well, that was as far as the train goes, we had to stop, that was in this little town by the name of Hosmer, but Dad was lucky enough, one of his friends from Russia was there and picked us off the train, and hauled us out to his farm. Mr. Meyer took us from Hosmer up to North Dakota with the wagon, three horses on the wagon. There was two families on that wagon, our family, and my father's brother and his family. There were 17 persons on the wagon, hills up, we mostly walked and hill down we all have a ride.

We were 8 days on the road, from Hosmer 'til we come to North Dakota. We stopped at the Coal Mine Lake and there we had dinner and Mother scratched every crumb together and there was no more chance to have a dinner or lunch, it was all used up. Mother said, "What are we going to do if we don't get to the grandparents by evening, we got nothing to eat anymore", and Father said, "We got so far and we can make it". So after dinner we start, we stopped at the first farm by the name of Goll. They asked him how far we got to go to the Feickerts [grandparents], and the girl told us that was four miles, and they were waiting for us yesterday already, so only four miles! That was one o'clock in the afternoon so that gave us plenty of time to get to Feickerts. When Grandma saw the wagon coming, about a mile or half a mile, she came over to meet us.

On Easter Monday Grandma was well prepared with Easter eggs, Kucken and milek (milk), and that was our first celebration we had in America, in North Dakota. Grandma was well surprised that the whole bunch came on Easter Monday.

Jacob Dockter, 95, Sheridan County

I came from Barasina, Russia. I left Russia and came to New York, and from there I came by train to Kulm, North Dakota. We were on one farm for just a year, that farm was too small for us. Dad had gotten more horses and more machinery and we moved to a bigger Farm. I don't know how much land there was. I was only two years old when I left Russia.

What was the reason they left Russia?

They heard in America you could take up land and you could farm here. In Russia they could only have a few acres, that is all the whole family had, and here you had hundreds of acres if you wanted them, and that's the reason they came over. There was no freedom at all in Russia. The Russian government had the thumb on you, you do as they say. That was the biggest reason why they left.

How was your trip financed?

My uncle borrowed it from the bank and when they came over, they signed up on the bank. Uncle John, my dad's brother, was here in Kulm. In Russia you couldn't make a living. Here it was easy to make a living—that's what we heard and that's what we followed.

Lucas Moldenhauer, 87, Dickey County

Where were your parents and grandparents born?

They came over here from Alt Friendental, Russia, but originally they had immigrated from Germany to Russia. They came over to the United States in 1902. They didn't know the English language, or had ever traveled and yet they got through the immigration deal. They found their way around and came to South Dakota first. Then they came to North Dakota where my dad's sister had previously come. They stayed with them until they could get a homestead of their own.

Edna Frank, 74, Sheridan County

My parents came from Russia where Mother's parents were German landowners who had Russian serfs working in the vineyards. Mother, as a very young girl, also worked in the fields. Her schooling consisted of a 2nd grade education. Dad and Mother were married when she was 19 and he was 27. She was probably 21 or 22 years old when they left their homeland for a land more free. Her parents did not wish to see their daughter go so far away, but she was determined. The trip over was hazardous. The first ship ran into a sand bar near Copenhagen, Denmark. The next leg of the journey was way-laid in Liverpool, England, where they stayed in a cheap hotel overriden with bedbugs. The seas were rough and only a few hearty souls remained "on to". Mother became very seasick and their 1½ year old daughter became very ill. When the ship arrived in Quebec they were quarantined until the baby either got better or died. Dad washed clothes in strong lye water, and acted as an interpreter to make money to pay the bills. The baby died and then they came to North Dakota via train.

Ella Guenther, 67, Dunn County

There are four sets of great-grandparents, and all of them came from Russia at various times. My father, grandfather, and grandmother came in 1884. My grandfather was born on the ocean, during the crossing. My father's mother's parents came in 1884 and my grandmother was born in Mento, South Dakota, on the way. It wasn't a state yet, it was Dakota Territory. They came into the northern part of the Dakota Territory, into the Zeeland area. My mother's mother was seven years old when she came with her parents and that was 1896. And then my mother's father came when he was about 14 years old and that was in 1906. Now the first three I mentioned came from Kassel, south Russia. My mother's father came from Klynordof, south Russia.

When they came to Dakota Territory, the good land in the Red River Valley was already taken up, the good land in the northwest part of the state was pretty well taken up, and the northeast corner was pretty well taken up. So we have what is called the German Russian triangle, which stretches north from Hettinger County to Rugby, and then southeast again to the Wahpeton area. In that particular part of North Dakota, when you get down into McIntosh County, where I grew up, there are more stones per square inch than anywhere in the world. The joke is told about the German-Russian woman who dragged her sauerkraut rock all the way from Russia because she thought she might not find any stones in North Dakota. I think this possibly could be true. They were a little disappointed when they came across all those rocks. They cleared the rocks and cleared the land, and were satisfied with what they were able to accomplish on the land.

Why did they come to the United States instead of staying in Russia?

Well, that is a long drawn out story. Perhaps I should explain why they left Germany in the first place, and then go on to why they left Russia. They left Germany in the 1800's. There were various groups that left Germany. In 1763 Catherine the Great, who was at that time Czarina of Russia, issued a Manifesto inviting Germans to come settle in her country, and the area she wanted them to settle was near the Volga River. She needed the people to show the Russians how to farm. She invited these people to come with the idea they were to have free land, and freedom from taxes, no military service, freedom of religion, and at that time there were religious wars going on in Germany.

Karen Retzlaff, 39, Nelson County

I was born in Argentina in the village of Macachin in the province of Buenos Aires. I was born June 1, 1912. My parents were German-Russian — of German descent. They were born in Russia and migrated to Argentina.

How come your parents moved to Argentina?

They weren't satisfied with things as they were turning out, and they were fortunate they left when they did. A lot of our ancestors moved to America then too, they saw the unrest coming, they lost privileges right along.

When did your parents first come to America?

They boarded the steamship Van Dike, a British Liner. We left Buenos Aires on April 12, 1914, and landed in the New York Harbor on May 12, 1914. It was a combination freighter and pleasure cruiser. They stopped at different harbors to unload wheat and meat ftom Argentina and some places they loaded fruit there. They went sightseeing on a guided tour, so to them it was very interesting. When they got to New York they stayed on the ship the first day, and then they had a place right there where they kept them. From there, they boarded the train for Nebraska. They lived down there for about two years and then came to Dickinson. They lived there for about a year and a half. From there they moved north of Gladstone, they lived there until the fall of 1925. They moved 25 miles south of Belfield — their mailing address was Midway, then Amidon.

My parents became citizens in April of 1927. I was in the eighth grade at the time — 14 years old. They had applied for citizenship in Dickinson and there was a waiting period, I think, of two or three years and they had their last hearing in Amidon. It was a legal process. They had two hearings before and were under surveillance. We children became citizens because we were under age.

What did they have to do to become citizens?

They had to answer questions, and they had to pay allegiance to their new country. They renounce loyalty to their own country. They had to learn the language first. They spoke German, Spanish, and Russian.

Bertha Obrigewitch, 74, Stark-Billings County

Freedom means different things to everybody. To those who have always lived with it, freedom may become very commonplace. But to people who have not always had that privilege, freedom is accepted as a gift. Born in 1904 in Zhitomir, Russia of German descent, all we knew was agriculture. That was during the reign of Czar Nicholas and Czarine Alexandra. Years earlier, the Czar had invited German farmers to come and teach his people a better way of farming. Many immigrated to Russia. The Czar promised that in return for their services, they would receive their own land and farms, a promise that went no further. The food produced by the villagers was collected by the government. The troops never left sufficient food for the families. We milked many cows, but we never had any butter — they took all our cream, but the family had plenty of sour milk with which to make various foods.

When the Bolsheviks' revolution broke out after the assassination of the Czar, the country was in turmoil. The elders in the village warned us not to say anything bad about anyone because someone might hear and things would not go well for us.

The little freedom that the villagers possessed was soon taken away from them. We couldn't worship anymore, the government took all of our food and clashed with the troops. People started to run for their lives, some fled on foot, while others escaped on trains and in wagons. We were so afraid of being caught that we traveled only at night in wagons.

My mother came from a family of five, but there was only enough food in the family to feed three of them. My mother and her sister were placed with different families who could provide food for another mouth. The two sisters did not see each other again until the Kolm family came to Moscow. The family received asylum at the home of an aunt and uncle. They completed their escape with monetary support from my uncle in Winnipeg. The family lived with the idea that freedom in the United States was the ultimate thing to possess. We had heard that things were so good in the United States that the streets were even paved with gold.

Because the immigration rules were more involved during 1913 in the United States, the family had to land in Quebec. As they passed through Customs the family was examined by doctors. As a result we had to stay behind because my brother had contracted an eye disease. While our other relatives went ahead, we had to stay behind I just screamed. I thought for sure we'd get sent back on the next ship. As soon as the eye inflamation cleared, the Kohn's were able to continue on to Winnipeg.

Coming to a new land was full of novel experiences. We were each given a large orange and a box of raisins. I thought the orange was so beautiful, but then I did eat it. I didn't like it because I had never tasted anything like it before.

I'll never forget my childhood experiences. Fear and want are not easily forgotten.

Edith Koehn, Cavalier County

Ayr was a small town that was four miles from our home farm. We went there to church and things of that kind and for shopping. The first settlers came to the area in 1879, but the area was really settled when the Scotch Syndicate, formed by a group from Edinburg and Ayr, Scotland, bought a land grant of 16 sections west of the resent site of Ayr. They called their land holdings the Park Farms. These sections of land were scattered throughout the township, but most of them were not too far from the home farm. Flemming was the first suggested name, then Elgin was picked. But because there was another Elgin on the railroad line, the name was changed to Ayr in 1884. F. H. Dickinson was given credit for that name. James Dunlap, one of these Scotsmen, was the first manager of this farm. Mr. Park was here at least part of the time.

The Syndicate sent stationary steam engines from Scotland to break the land, and the plow was pulled by cable between two engines placed a half mile apart. They kept two teams of men, most of them had come from Scotland. With their six plows they planned to break a quarter section per day. They had to send back to Scotland for repairs, which must have been rather slow. The engines were moved by horses and mules. An important man on the crew was the block setter whose duty was to even up the cable on uneven ground. Some of this cable was found on one of the surrounding arms. This area was a point of interest and gathering spot for miles around, and people came by carriage or horseback from Casselton, Tower City, and other towns for special events on Sunday afternoons. The original house still stood until a few years ago, although F. H. Dickinson improved it. Other smaller homes on the farm were built for the crews. They owned over a hundred mules and a fine drove of horses. When the cable was no longer used, they were shipped to a grape ranch in California, and that was about 1898. Thomas Punton came from Scotland to work this cable on the farm. (Two years later he became manager and spent 10 years there before buying the Elgin farm of three sections. This farm had been organized in 1883. The Park farm donated land for the village site.

After the railroad came through, Cargill Grain Company built a "flat house" at Ayr a year later. Blind horses were used for power in the elevators and also for the first threshing machines. Early harvesting was done with a cradle. Before the elevators were finished, grain was hauled to Wheatland. She [mother] also told us that wheat was hauled from Mayville to Wheatland by the early settlers because there was a "flat house" at Wheatland.

What was a "flat house"?

It was for grain. I don't know exactly how it was constructed. It wasn't the tall elevator. My grandmother was very generous, and so many nights people from Mayville, either going or coming to Wheatland, would stop there and feed their horses and water them. And she would feed them [the men] and I suppose they would sleep in the shed in the barn or somewhere. I can remember Mother telling us her father had been in the gold rush in California in 1850 and had made a little money. When they moved from Canada to here, hc showed his children $12,000 in gold coins, telling them that they would probably never see that much money again. And like I say, he eventually lost his farm.

Ellen Cameron, 83, Cass County

My grandfather, my dad's father, came from Scotland, his cousin was Alexander Graham Bell. When the homestead land opened up in western North Dakota we came out here between Christmas and New Year's in 1906 and filed on a homestead. And he said, "Well, just so I have a quit claim to my farm." "Well," he said, "It's OK". But there was snow on the ground and no one could tell what it was like. So we moved out there on the first of April 1907. Lloyd, my brother, and my dad came on an emigrant car with horses and cattle and furniture. My mother and we four kids came on the train. We got into Richardton at 2 a.m. in the morning and there was no one to meet us, so my mother went to the local hotel and we four kids were trailing behind her. She went to the clerk and said she wanted to hire a liveryman to take us out to Frank Moffit's. He said, "I'll tell you, lady, if I get a man up at this time of night he'll charge you $5.00, and they only live up here about a half mile. I'll light a lantern and you can walk out there." So that's what we did. And as we were walking out there, Bill [brother], was three years old, and he said, "Ma, is we walking out West?" I 'spose that's all he heard from the time my dad was out here to homestead until we came.

Lillian Krieger, 87, Dunn County

When my grandparents emigrated from Sweden in 1855, they had a long hard trip across the ocean. The wind was strong and the sailboat went backward as much as forward. The trip took six weeks. Grandma was seasick and someone gave her a herring. She kept holding the herring between her teeth. Their two small children died from lack of liquids, I guess they were dehydrated. They were buried at sea. Grandma mourned the loss of her babies.

Ruth Stromberg, 91, Cass County

I was born on January 25, 1911 about 8 miles north of Killdeer in a log house. Dad's parents came from Illinois, Mother came from Anoka, Minnesota. They came to make a new home in the west. His wife (Grandma) was sickly and he thought this climate was better.

Are you of English heritage?

No, I'm like a mutt dog, the best of all breeds.

Verne Howard, 85, Dunn County

The folks lived on squatted land here, land was not surveyed when they came up here.

What was squatted land?

You could settle on any land then that was not occupied. When my folks came here their closest neighbor was a distance of 15 miles across country. The place where they settled here was on a hill, there was no wind — if the wind came from the west, no wind could strike the cattle. They fed along the lake edge.

Isaac Allen, 90, Sheridan County

They came from the East and they didn't know much about farming. They were certainly brave to come to North Dakota, but you know the land was free. The Saumurs came from Ohio to St. Paul. They had been used to working in the shoe factories in New York state. The Gotzian Shoe Company opened up a shop in St. Paul and any of their employees that wanted to go to St. Paul, they would bring them there. Well, the Saumurs came and worked until the land — claims — opened up in North Dakota. And they had enough money to buy a wagon and a team of horses. And their oldest child was a baby. She (Mrs. Saumur) held that baby on her lap all the way.

Marie Saumur, 94, Grand Forks County

When did you come from Iowa?

In 1914. Olaf had an uncle that lived in North Dakota and he wanted him to come up here. His [uncle's] health was failing and he wanted us to live there. Well, Olaf came home and we talked it over and decided that it would be all right. So then we had a sale and sold all our horses and cattle and we shipped some household belongings to North Dakota.

Uncle Andrew, the one that we bought the farm from, came to Hankinson with a lumber wagon and we rode in that. It was a very nice day, and I sure enjoyed the trip to our home in North Dakota.

Clara Anderson, 93, Richland County

I was born in Glendon, Iowa. My folks moved to North Dakota in 1902 where they took a homestead — right close to the Truax Coal Mines. My father hauled coal for a little money until he could get some land broken and get a crop. His first crop was 10 acres. Of course that didn't bring him very much money. My mother was on the farm for three years without ever stepping off the farm.

Nellie Erickson, 86, Ward County

There were people who came from the east in order to find homesteads. They had to live on the homestead 3-4 years and raise crops and they could then prove up on it, and sell it at a good price, and they would then go back to where they came from. For some it worked, but for some it didn't work at all. Those people who came from Iowa, they were nice to get along with, and the men would help each other in the fields with their horses, and we liked them people. They were nice.

Marion Schmidt, 84, Stark-Billings County

I was born in 1899 in Laverne, Minnesota, the oldest of a family of 10. Dad was farming, and you couldn't get enough land. He only had one quarter and couldn't make ends meet. He went to the bank to get some money to get some land and the banker said, "Well, they give away land in North Dakota, why don't you go up and see what you can do about that?" So, he came to North Dakota, came to Minot, planning to go to Mohall to get land, but a big storm came up, this was in March, 1902. He had a friend in Berthold, so he came up there to come through the storm and didn't it last three days! And so Fred Fink, this friend of his, said, "Well, I know of a quarter right over here just a little ways that's for sale because the man is gonna give it up. It has a nice house on it." He would be able to buy that for $300. — a relinquishment they called it. Well, it had a two story house.

Dad went back to Minnesota and brought his horses, cows and machinery on a box car. He got settled on the place and sent for my mother and we two children, I was two years old then. The day we came, I have been told, my mother was met by this same man, Fred Fink, because my dad couldn't come. The horses had run away and they'd gotten clear up to Donnybrook before he found them, and he was on that search. Well, this friend brought her home and put her in this little house with the stovepipe up through the ceiling, rain had run down and it was rusty. She just sat down and cried and she said, "If I'd had a return ticket I would have gone back to Minnesota." But of course she stuck it out.

Florence Smith, 87, Ward County

I was born at Willmar, Minnesota in 1900, and my dad had a music store there. My brother passed away in 1902 and so did my dad. So my mother took us, I had two sisters, and she went to North Dakota. My grandmother and her two daughters and her son all had moved up here and homesteaded. We came up here to stay with them 'til we could get on our feet again.

When did your grandmother come here?

1899, I think.

She must have been one of the early homesteaders then?

She was. She didn't know she had neighbors until one morning she heard some pounding. Her son got on a horse and started off in that direction where they heard the pounding, and found out there was a man building a house. When she started living on her homestead she was scared because the Indians used to go through there a lot. She had a well and they used to stop there to water their horses. One time her youngest daughter was out herding a cow and a calf and the Indians came and she hid in the tall grass, but the Indians did see her and one of them came over. I don't know if he could speak English or not, but he made her understand, anyway, that he wasn't going to hurt her. Then he gave her a pony. She thought a lot of that pony, she kept that pony 'til he died of old age.

Minnie Ness, 82, Burke County

Mother tells about her girlhood days in Minnesota about the time when Jesse James was making headlines. She tells of four men and one woman coming to their home and wanting to stay overnight, they were willing to sleep in the barn. Grandpa said, "Well, yes, they could", but he was very leary about them. They did stay in the barn, but Grandpa worried all night long for fear his horses would be gone the next morning.

Lena Allen, 85, Sheridan County

In 1907 Aunt Sallie, who had just graduated from business college, was hired as a stenographer by Will Clarke, manager of the Gold-Stabeck Land and Credit Company at Flaxton and Bowbells. She later married him. In 1910 he was transferred to Dickinson. After trailing a deer while hunting in the Badlands, he staked a claim north of the Killdeer Mountains in Dunn County. He then had Aunt Sallie and her sister, Aunt Eva, file on adjoining claims. Aunt Eva had come to Bowbells in 1908 to work as a registered nurse under "Doc" Paulson, so she spent the summers on her claim and winters in Bowbells.

In 1911 Aunt Bessie and my mother, Lillie, took the train to Dickinson and came overland the hundred miles in a stage and buckboard to Aunt Sallie's, fording the Little Missouri in the process. How in the world anyone found the way to their destination through the Badlands, I'll never understand.

Arlene Schultz, McKenzie County

They [Grandparents] had two immigrant cars, one for the cattle and another one for furniture. They were caught in a March blizzard, the train stalled out of Kenmare and was buried in snow and the cattle all smothered. The Soo Line paid them $16. a head for the cattle. Then Grandfather went back to Minneapolis and bought a lot of Black Angus — the first Black Angus in the county. And, by the way, we still raise Black Angus.

Ruby Bryan, 73, Burke County

My dad homesteaded 12 miles northwest of Anamoose, and this was where most of our family were born. When my folks came to this area they came in a covered wagon pulled by horses. My mother had gotten a new sewing machine from her folks, which she brought along, and a cow that was tied behind the wagon. Here they built a sod house, dug a well, and built a barn. We lived in this house until we bought a different farm.

Mabel Albrecht, 80, Wells County

My father came and bought land and built a house in 1899 when North Dakota was still a very young state.

Was that before the railroads came through?

It was about the time the railroad was in the process of coming in. Jim Hill had made plans for the railroad to come through Wamduska — south of Lakota, then he changed his mind and put it through Lakota and on west to Devils Lake.

Can you tell us a little about Wamduska and what they did to prepare for the coming of the railroad?

They built a large hotel and it was made from brick from North Dakota clay. It was on the banks of what we call Stump Lake. I don't remember how large it was but the dining room was very large. One could turn around inside it with a team of horses hitched to a wagon. It was never used as a hotel, but part of it was rented out and it was used as a recreation hall or community center.

Mary Mootz, Nelson County

My husband's home was by Thompson Lake and he had lived there since 1900. His mother had come to Bowbells in 1900 in April. Her husband had passed away in Minnesota in March and then she came to North Dakota with seven kids, the oldest one was 10 in June and the baby was 17 months. She homesteaded out in the hills. She came up from Minnesota with an immigrant car. For a while she had two oxen, and then she finally got horses, and she brought a cow from there. It wasn't very good going. Then they started a rural star route for mail from Kenmare in 1909. Then she hauled mail three days a week and that was 18 miles one way, and she got $250. for a year.

For a year?

For a year and that was with a horse and buggy in the summer and a team and sled in the winter time.

Margaret Lien, 69, Burke County

My dad came in 1898. He wasn't married yet then, but he came and then he went back to Minnesota. He had a threshing machine and a well drilling outfit. I guess he went back there and threshed and then must have sold that. Then he brought the well machine and his horses and a cow and I think four hens and a rooster.

He came by train then in an immigrant car?

Yes, immigrant car and he had some machinery.

Myrtle Bush, 82, Bottineau County

In February 1906 my dad moved seven horses, some machinery, and all their household goods by rail in one boxcar. Then, a week later my mother and baby came by train.

There was quite a few farms built up here by the time they came to North Dakota, but there were some homestead shacks too. This was in 1906 and most land had been homesteaded by then. My mother told how they were here only a couple of days and they were invited to a neighbors home for a party, so they met and got acquainted with people quite soon. Their farm was only four miles from town. The town was fairly new then too, Goodrich was established in 1901.

Irene Sauter, 69, Sheridan County

In the early pioneer days, mornings were so peaceful. The roosters would start to crow at daybreak. Seems like there was dew on the grass every morning. I don't recall we ever had a dry summer, it rained often. Apple, plum and cherry trees blossomed in the spring and the air smelled like cologne. Hummingbirds with their long beaks would reach in the sweetpeas and morning glories to suck the honey out.

Ruth Stromberg, 91, Cass County

I shipped an immigrant car up from Missouri.

What all was in that car? Can you describe an immigrant car?

It was entirely for your personal belongings that you want to move, I had also bought some new machinery and some horses, 500 fence posts, and I think it was 500 feet of oak lumber oak and hickory lumber that I'd gotten in Missouri.

Isaac Allen, 90, Sheridan County

I was born in Missouri in 1916 and came to North Dakota in 1921. We came to North Dakota when my Uncle Denny Hallahan passed away and we came to take over his homestead. He was a soldier at Fort Yates, North Dakota at the time of Sitting Bull and Custer.

What war was it?

It really wasn't a war, only the Sitting Bull uprising to help protect the pioneers from the Indians.

Was that Dakota Territory?

Yes, that was before North and South Dakota were separated, there was no Grant County, it was all Morton County, now it's Morton, Grant and Sioux. When Uncle Denny passed away my daddy decided North Dakota was a much healthier state than Missouri, so he loaded a railroad immigrant car with our family dog, Rover, and as Daddy was in the mule business in Missouri he loaded a "Jack" for a Shields rancher, Mamma's 1886 "New Jewel" sewing machine, harness, saddle, tools and many other useful things that we needed here in this pioneer state. We took a big wooden box full of Mamma's quilts, and as she was packing them I was busy hiding dolls under each quilt. When we got to North Dakota it was March and the wind was blowing, lots of snow, and I can see Mamma yet when we got off that train, that sad look, 'cause she had come from a better life in Missouri where there was telephone service and close relatives.

Carrie Weinhandl, 70, Grant County

What prompted your family to come to North Dakota?

Someone came down and told about the big North Dakota crops — 40 bu. to the acre. Dad thought things sounded so good. He had money and three farms in Nebraska. So he sold two farms and came to North Dakota and bought nine quarters of land.

Where were your parents and grandparents from?

Mom's parents — The Claussens, were from Germany and could speak no English. Dad's parents — The Arthurs were from Wales. They homesteaded in Nebraska. They were all farmers.

When we moved, Dad shipped his car, a Model T Ford, some machinery, horses and cattle up from Nebraska by train. He came up about a week before we did. It was storming when we came so we had to stay at the hotel for two days, it was the 20th of March. Dad came for us with the horses. People thought he was crazy taking us out in such bad weather. When we got home he told Mom not to let any of us out of the house.

Viola Close, 77, Towner County

I was born February 12,1904 at Craig, Nebraska. My parents moved a month later in an immigrant car with several families to Tagus. We settled in a tar papershack on a 160 acre farm, Dad had purchased this claim relinquishment. There were four children who were ill with whooping cough. We had no relatives here and the neighbors were far and few. All were Scandinavian, we were of Irish and English lineage. Over the years we blended socially with them or they with us.

Golda Reid, 81, Mountrail County

I would have to go back to when my Dad left Pennsylvania with my brother and I. He traveled all over the United States and had us two children with him and when he would get to a town or a place, he would look up the Salvation Army or the Ministers. Back in those days they didn't have welfare and stuff like they have now — so they went to people for help and either you were given something or you were turned down. Well, he traveled in all the states — I was in all the 48 states excepting five by the time I was seven years old. When we got to Aurora, Nebraska, he entered the Salvation Army there and he asked for help like he did other places and he told them a story, and I always was the type that would speak up the way it was.

I really should have started out when my mother left my dad with four little children and went off with somebody else. The youngest one passed away when she was 11 months old. My aunt took my one brother, and my dad took my other brother and I and traveled with us. When we got to this place I told them our mother wasn't dead, that she had run off with another man, then they took us to a courtroom. I remember so plainly, we were sitting in this courtroom and there was this judge and the Salvation Army people, and they asked all kinds of questions. That afternoon they took us to Lincoln to the State Orphanage and that's where we were placed, but I was adopted out in March and I grew up at Henderson, Nebraska.

Mabel Adrian, 73, Cavalier County

I came to Bowman with my mother and my six brothers and sisters on the train, and my father and my oldest brother came in an immigrant car. We came from LaFarge, Wisconsin, and we arrived in Bowman just as the sun was setting. My grandfather met us with the lumber wagon and horses and took us out to his place northwest of Bowman. We were very much surprised with there being no trees.

Vera Storer, 85, Bowman County

When I was about five years old my father decided that he wanted to take up a claim out west. He had bought a claim in South Dakota where we lived before we decided to go out west. Then we had to quick get things together to go out in the open country where there were no trees and no neighbors. We had to get out there by loading what little furniture and things we needed away from towns or anything, and we had to load that all in a boxcar. They had to haul everything up to Fairmount and load it in a boxcar. We had a walking plow because you had to break up a certain amount of land, and this land wasn't really so much for farming. We had to use so much of it for building sod sheds.

We had a double surrey, an open buggy, with a shade on top with all the tassles, and we had two horses, Prince and Jenny. They had these light flowing tails — pretty horses, and we had one cow. I don't remember the cows name at all. We only had one so we should have found a name, and we had a few chickens. There was very little furniture, just some plain beds. We had an ice box, and that's where we stored the stuff in when it was cool, it was out on the side of the house. We got everything that we could take along in a boxcar and then we said goodbye to the horse and the cow and we rode in the caboose — a little place where the people rode on the end of the train. Everytime the train stopped for any length of time at some place, then we would have to run out and see how the horse and the cow were making the trip, then we would quick get back on. When we got out there, the boxcar was pulled to a siding on the track so we could get out.

Well, then we had no buildings there, just nothing but open prairie and kind of tall prairie grass. We had to have a place to stay and there were some friends from back in South Dakota that had gone out west a couple of years before. They lived three or four miles from us, kind of near a little river. So the dad met Mother and us little kids and we stayed there overnight 'til he could get back and start hauling the stuff from the track. The first thing he took was a load of lumber with some of us, and we dug into the side of a hill. We dug down just enough so we could stand up some straight boards along the front and then slanting boards for a roof. They piled kind of sod on top of it to hold it. Then we had to get the stuff from the box car out to the farm and set up beds.

Tell us about Hulda, your little sister.

Oh yes, we had quite a scare. We had no regular roads and once in a while we could see cowboys going on horseback, galloping across the land. We didn't pay too much attention to that, but one day, after my dad had started out for town with the lumber wagon to get more lumber and supplies, then all of a sudden, Mother started looking for my little sister Hulda, the one that could just barely run around. There was just the shed where our horses and cows lived and our little square house and we looked and looked and we couldn't find her, and Mother got so very excited. We saw some cowboys riding across the hills and prairie, so she quick grabbed the string of her apron. She took that apron and she ran hollering and waving the apron up in the air to attract those cowboys. They came galloping up with their big wide colored chaps on their legs, they wore that so their legs didn't get scratched from the bushes or the tall grass that was there. Then she said, "Oh, will you go look for we lost my little baby and she can just barely walk. Maybe she tried to follow the lumber wagon when Dad went to town." So they quickly tried to follow the track back to town. They started out and then they didn't find anything. Then they came back and they started galloping around, kind of through some bunches of grass. Then suddenly someone happened to look and there was one window in our first little shed where we kept the horses and cow, and here she had gotten tired and crawled down in there and she was sound asleep. We were so thankful because we had no idea where she could be and it would soon be night.

Ethel Pirmantgen, 86, Richland County

When we first came here there was no Rhame here, no Bowman, no Marmarth, it wasn't even surveyed then. They didn't survey this country 'til 1905. It was the entire Billings County then.

What made your family come out here?

My people were in the Jackson County War in Wyoming. The ranchers ran them out and they came here and settled — they just took squatters' rights. It's in the history books — it was the big ranchers at Rock Springs, Wyoming. They ran all of the smaller ranchers out of there. They didn't want to be surrounded, and my people were some of the smaller ones, so they were the ones that had to leave. I had an uncle that they killed at Osage, Wyoming — and they wouldn't let my grandparents bury that boy. He's buried at Osage, but we don't know where.

Hilda Putney, 91, Bowman County

My parents took out a claim near Ortonville, South Dakota and lived there for two years. In the fall they went back to Wisconsin and sold the homestead there and the family went to live with one of Mother's uncles until they could get settled. So loading up some machinery, horses, and a few cows, they headed for their Dakota Territory homestead once more. Having spent more time there in South Dakota, they were not too happy, so they decided to head out further north. In 1880 they loaded up their belongings and headed north over the prairies. Most of the pioneers headed for rivers and woods. That's how Grandpa Iver and Uncle John came to the Sheyenne River Valley as there they could cut trees into logs for their cabins, firewood, etc. The river water and sloughs were clear fresh water and were used until they could get wells dug.

Alice Stahl, 75, Nelson County

Our farm was homesteaded in 1892 by David's father, who was Ernest Fairbrother. He came up from the Red River Valley, having come to Forest River from Wisconsin. Being dissatisfied there, he came to this area and homesteaded on the quarter of land on which the home is now. He lived on it himself until 1941. We are just a few years away from being a Centennial farm. We hope that by the time it reaches the 100 years that we will still be living on it and enjoying doing so. We will perhaps have a little plaque or do something that will designate it as 100 years old.

Vera Fairbrother, 71, McHenry County

The land here was settled so my dad had to contest the place where we were supposed to move on. We had to wait a year to see if that man that had it would come back, before my dad could contest it. We had a sod stable the first year. My mother and father built that.

Cora Lykken, 93, Nelson County

Some years ago a friend wrote a history of her family and told about some of the things they experienced in their pioneering years. The family had settled on a claim south of Fillmore in Benson County. When they first came there they met another family who had come to the area some years earlier. The woman said she and her husband had lived there eight years before she saw another woman.

Anna Bullock, 72, Pierce County

My grandfather homesteaded the farm in 1864. I should have a copy of that here some place. His wife died before he came to North Dakota to homestead. He had a son and four daughters. He got tired of farming without a wife so he left the farm and sold it to my father for $500. when my father was 19 years old. I think he only had two oxen, a plow and a few other small things to start out.

Merlie Cruff, 75, Barnes County

I was born in Texas and as a young girl I picked cotton all day long and crawled on my knees and put it in a sack. This sack was strapped over your shoulders and you would drag it behind you until it was full. Then you would weigh in and put it in a big wagon and then go back to your knees again and start picking. I used to work in a laundry when I was young and I made 17¢ per hour.

Edna Klettke, Dickey County

They came to Strasburg with his folks and settled there. He was raised there. Then they got to know some folks that told them about Wells county, so they came here. They homesteaded near Bowdon because they thought the grass looked greener. They found out later it was sandy soil.

Leona Kuske, 83, Wells County

You mentioned that you lived on the ranch until you were about ten years old. Where did you move to then?

It was so hard to get a school close by, then some people were up to visit and they said, "Say, there is 160 acres real close to Hebron for sale. Why don't you inquire about that?" So my dad came down and he made a deal with them, he bought the 160 acres. There were so many homesteaders coming in at that time and taking claims, so they had no trouble to sell the land and then a neighbor bought the cattle. An uncle came and helped with moving and they had two hay wagons. They loaded all things on the wagons, and my mother, brother and I came with a neighbor to Dickinson and then we took the train to Hebron and stayed at Grandma's house until Dad came with the things. First we lived in a tent, then we built a grainary and lived in that until the house and the barn were built.

Rose Schmalenberger, 84, Morton County

Our parents land was separated from the neighbors by the section line. It was also grazing land on agreement, we herded cows on the section line. It was also a thoroughfare used by the big lease ranchers, chiefly the Bill Connolly Ranch, to herd the cattle through on the way from the reservation badlands pasture to the main railway shipping point of Killdeer. It was a delight to watch the herd trailed through with cowboys at the head, cowboys splitting the herd midsections and cowboys trailing behind. It was quite a line and always a wonderment.

The Indians from the Reservation used the section line also when they traveled to Killdeer or other points, they used lumber wagons drawn by horses. This continued until about World War II or later. I believe the building of Garrison Dam brought an end to that era and the Indian people also began using automobiles. Roads were improved too.

Ella Guenther, 67, Dunn County

One night in Tuttle, Soren Sorenson, better known as "Shorty" asked me for a ride home. When we got there I had to come in and taste his homemade Juneberry wine. He was in a talking mood and I was willing to listen, so this is what he told me. "I knew your granddad, I say he was the strongest man that ever came to Tuttle. He was unloading flour from a box car. When he heard the train whistle, he lifted his wagon with several sacks of flour on it and set it down away from the car. Your grandpa didn't know anything about fighting but he was just so strong you couldn't do anything with him.

I asked him, "Did you know Ed Counts?" "Yes," he replied, "That was Jack de Rustler". He also had an underground barn on the south side of that long hill. When I was riding for Tommy Lloyd who ranched near Woodhouse Lake, Tommy said to me one morning, "Jack de Rustler has one of my colts, I want you to go and bring him home". So I rode over and told Jack that I was taking Tommy's colt. Jack answered, "If you take that colt I will get Tommy Lloyd". Sure enough, a few days later the sheriff notified Tommy to answer a complaint for horse stealing, so we had to take off for Steele. Tommy could drink and play cards during the day but the sheriff had to lock him up at night. When Jack was asked how he knew this was his colt he answered, "There is a dime in the colt's shoulder." The dime in the shoulder was a common way to cure a horse that had been sweenied, it prevented the hide from growing to the shoulder blade. I also heard parts of this story from my mother. It still shows where the underground barn was. Jack de Rustler was later shot in the back by one of his own men, near the Canadian border.

Edwin Johnson, 71, Kidder County

At the ranch we didn't have any neighbors — you'd have to go eight or ten miles to get to them. It was cattle country.

Did you ever see any Indians around here?

No, I don't know who started that tale. They seemed to think that Indians lived here, but the Indians were gone when we came. They used to go through here — maybe going to some other tribe and they followed the river. They weren't hostile Indians. They camped along the river there, but they never bothered us.

Hilda Putney, 91, Bowman County

My father's mother lived in Fort Yates. She was a Hidatsa woman. The Hidatsas now live around New Town. She came to Fort Yates in the late 1880's to attend a big celebration there. She came down with her brothers, the Packincays who were famous interpreters for not only our people, but their people. Her father was a fur trader. She told how he had taken her to school down in St. Louis, but she cried so much that they brought her back by boat up the Missouri after just a few months. When she came to Fort

Yates, my Grandfather Defender wanted to marry her. Afterwards she said her brothers left her and she was so lonesome. Although she spoke Hidatsain she didn't speak English and she didn't speak the Dakota languages. When Defender died she married one of his cousins, a man named Pretends Eagle. She had a son, my uncle, his name was Dan Pretends Eagle. When Pretends Eagle died she said her brothers came after her but she told them, "No, you brought me down here and left me so many years ago, now I'm going to stay here until I die". Then she married Red Pheasant, Red Pheasant was a policeman. They lived at Fort Yates so that is where we used to go to visit. My mother would take care of whatever business she had to do at the Bureau of Indian Affairs office. We'd stay there probably one whole day and then start back the third day.

The Dakota women had many responsibilities which have not been recorded because in those times people thought it was pretty much a man's world. It certainly was with the white people and they thought the Indian people were the same, but then we weren't. We were a very balanced culture where both sides contributed their own unique parts to what made life total.

I gather that my great-grandmother who died at about the time I was born did not deliver me because she was 90. That is why Annie Goudreau did. My great-grandmother was born back in 1840 and she lived with the Dakota people at the time, they lived in their area east of the Missouri River and lived in the times when they made all the famous treaties of 1868 (Fort Laramie Treaty) and the agreement of 1877. Part of the group that was forced to come to this side (west) of the Missouri River. Those were things that were told to us when we were growing up, her history as well as my grandfather's, [See the Bear's], history. On his side of history there are people who are written up in exploratory history going back to the 1700's. There were always people who were in some kind of leadership position whose name was See the Bear, and his people had knowledge to heal certain sicknesses. So on both sides of my mother's family we had people who did things.

Your father was an outstanding cowboy in his day.

My father, George Defender, Sr., I gather was an outstanding cowboy, I don't remember him as he died when I was about a year and a half. So I guess I always considered someone like See the Bear, more like my father, but we have photographs of my father and I remember seeing them as a child and hearing that he was an outstanding cowboy.

Mary Louise Defender Wilson, 56, Grant County

Lloyd and Lillian, my oldest brother and sister weren't very old, they would not have been over 11 or 12 years old. We got all of our wood down along the river, picked it up. They went down to get a load of wood one day and they saw a man washed up along the shore — a dead man. They ran home and told my dad about it and he went down, and then he got in touch with the authorities and they came and picked up the body. We heard afterward that he had been knocked on the head. They found his relatives, he had a daughter in South Dakota that came and claimed him.

Wasn't there a massacre there once?

That was before we moved to Dakota, just a few years before. That was the Spicer murder. That was across the river from Fort Yates, old Winona. Now there is nothing there any more. At one time it was quite a little town. There were several saloons in that town of Winona and these Indians came over there. There were settlers too, that country was pretty well settled and it was pretty good land there on the east side of the Missouri. These Indians came there to get liquor and this fellow refused to sell it to them. It was against the law to sell it to the Indians. To get rid of them, he said to the leader of them, "Go up north a couple of miles, you can get some liquor up there," never thinking that they would go. Oh, he must have lived with that the rest of his life! They were nice people, these Spicers, and so the Indians left and went up to this farm. There was a woman there with a pair of twins about a year old and I believe it was the husband's father, the grandfather, you know. Her husband had gone to Linton. They came there and demanded liquor and of course she didn't have any liquor, they weren't dealing in liquor, but they wouldn't take "no" for an answer and they got into a fight and oh, they had an awful time. The Indians killed the grandfather, she [Mrs. Spicer] fought them with the hoe, she put up an awful battle, but there were three or four of those Indians and one colored man. They took those twins by the feet and knocked their heads together, just knocked their brains out. They said she got the hoe and she chopped into them but then she got her hoe caught in the clothes line and then they killed her. I don't know if they did any scalping, they were shrewd and they left. That happened just before we came up here — about 35 miles from our place. Anyway, those Indians fled back across the river and that time there was a posse after them. They were caught and they were hanged. The Negro was proven to be innocent, he did not take part in that killing.

Hazel Winter, 75, Barnes County

Oh, I can remember how in the earlier days, every summer, a long wagon train of Indians would pass by our farm. There were about seven or eight teams, each with a wagon, sometimes extra horses tied along side or behind and also little colts would be following them. Then there were Indian men and women and children, they had boxes and blankets in the wagons. My mother said they were traveling from one reservation to another.

Irene Sauter, 69, Sheridan County

Many years ago I used to attend all the pow wows that were held at Shields and the Porcupine Sub Station. I always danced. I have many, many good friends, and when I was adopted into the Sioux Tribe, I was given a complete outfit. The headdress was given to me for a special occasion, and that headdress I gave to my boss during Recreation Days. We had a ceremony at the Sub and he was adopted into the tribe and he was presented with this headdress.

Bea Omber, 81, Grant County

In my father's native pasture land there was a circle of stones. He said the Indians had camped there and they had used the stones to sit on. Our neighter said the Indians had traveled through their yard on their way to Canada. She gave them fresh baked bread. That was before 1916. Indians were on the reservations in the late 1800's. They never bothered anyone.

Mary Hodgins, 76, Cavalier County

Grandpa would often talk to us about the Indians and tell about eating with them. Their meat was very good until he found out that it was skunk meat he had eaten. Grandpa said it didn't stay with him very long.

Alice Stahl, 75, Nelson County

The Indians used to go through our place on their way from Bullhead, South Dakota to Shields, to their pow-wows and to many other things. It was the Indian Trail right through our yard. They would stop and get a beef from us and butcher it right out on the prairie, then load up their meat and go on their way.

Gladys Morgan, Sioux County

Did you have any experience with the Indians?

With the Indians! The Indians were pretty much out of the country. They were not bothering anybody anymore.

In the evening I and my sister Mary went after the cows, to get them home for milking. We kids were playing, and when we came around the curve, here was an Indian starting to settle down for the evening, camping. When we saw those Indians we got so excited and instead of running home, we run a different direction, west instead of east. The cows came home, and I and my sister Mary were missing, and it was getting dark. The folks went out looking for us, and finally we came home, we must have made about seven miles, we were all pooped out. We were so scared we ran away from home when we saw the Indians.

Jacob Dockter, 95, Sheridan County

The Indians have a burial ground on one of our hilltops. They laid rocks in a circle to mark their graves. When our city friends came out they were determined to dig into the one grave. They figured that they were buried with their personal possessions and they just had to find out. They were buried in a sitting position. Dr. Arneson had the head of one, the teeth were in very good shape. He had that in his office for some time, then he gave it back and I think it was put back. We did not like for them to do this. I would have nightmares the Indians would be chasing me over the hill and I'd run for dear life. Night after night I would wake up all sweat and hot, but they never caught me. It was terrible and I think it was all because we had been bothering the grave.

Alice Stahl, 75, Nelson County

When I hear the word immigrant, my thoughts go back to my childhood days and the long ride we had on an immigrant train. Prior to that time, 1906, my family, the James Haney family, lived on a small farm near Glenwood, Iowa. All the farmers there had good houses to live in, nice houses, yards, and orchards, etc., but my father, who was then probably in his early 30's, got the homestead "bug". Mother said all she heard was "free land in North Dakota". But by 1906, really all the best land was taken. He and some other fellows came up to Dakota to file on a homestead, this was in Emmons County, about 25 or 30 miles northwest of Linton, the county seat. We were close to the Missouri River, just across the river from where the Cannonball River enters into the Missouri on the west side. The Cannonball Trading Post was about 2 miles south of the Cannonball River where there was also a government office to help take care of the Indian affairs. A doctor was stationed there too at that time.

Our immigrant train was made up at Council Bluffs, about 20 miles north of Glenwood. I don't remember how many families were in the group, but I do know that each family had a team of horses, a wagon and so forth, a milk cow or two. Some had a few chickens and perhaps a pig or two and a cat and dog. I remember one lady had a bird in a cage that she took with her in the coach. This was very interesting to all the youngsters. Of course every one brought their furniture, which was very plain in those days. The chairs were homemade benches, we sat on boxes or anything we could get to sit on.

We made several stops along the way to feed the stock and so forth. We landed first in Linton, where some were to settle and then we went on to Hazelton, a few miles north of Linton where there were some settlers that met us there and helped us move on to our destination. It was in early March, and as I remember, there was a Chinook wind and it was melting the snow very fast.

Well, to make a long story short, in a few days a tar paper shack was built. We used water from a spring close by 'til Father got a well dug much later in the summer. There were six of us children ranging in age from one year to ten. Five more were born in North Dakota. The neighbors were all very kind, they helped each other a great deal. Father broke up as much sod as he could with just a team of horses and a sulky plow and he planted some grain that spring, but it was a very, very dry year, and very little crop or garden was raised. So Father worked away from home to earn what money he could to provide for us. He made a dollar a day.

Why did anybody want to leave Iowa when it was developed and there were farms and nice homes?

Well, the farms there were very small and that was an old country. You see, they settled there shortly after the Civil War, came there in covered wagons and so forth. That land was all taken, about all you could get there was rented land. Then it was all the advertising that the railroad, the Northern Pacific, and the big railroads that went through Dakota, they were advertising these big land booms, and all you heard was the wheat up in the Red River Valley. To those people it just sounded like heaven on earth.

Did people get homesick, and did people go back?

Oh, did they get homesick! My mother, I don't know how she stood it. She had never lived in the country before — maybe in Denmark she did, when she was a little girl, but she lived in a city, Council Bluffs, all her life until she married Father. She was so afraid of the Indians, and I can hear those coyotes yelping yet.

Didn't she want to go back to Iowa?

Well, we had no money to go back, those immigrant trains didn't go back. They brought more up but they never went back.

What made up a homestead?

The government, they would give you — my father had 160 acres of land, but you had to live on it, I believe it was five years, cultivate the soil, and then at the end of five years, if you could prove that, have two witnesses to prove that you had lived on that homestead, and then the land would be yours. But in the meantime, he bought 40 acres adjoining there, that was rough land — didn't amount to anything. I don't know what he paid for it, I just don't remember, but it wasn't very much.

We lived on our homestead just a little over six years. My father proved up and he could see there was no way that he could make a living there. Then we moved to the west side of the river. There were more opportunities over there and more land and a chance to get into cattle there. We did much better over there. A lot of the homesteaders left and went elsewhere. They never went back to Iowa though.

Tell about the Indians that lived near by.

We lived by the Standing Rock Reservation, just two miles from us and just across the river — Missouri River. It was just a few years since the Custer massacre and they were pretty wild but they were starting to get tame. They were put on the reservation, teaching them to live like white people.

Who were some of the chiefs that were living there?

Well, I didn't meet any chiefs then, that was years later. When I was 17, I met old Chief John Grass. I do remember when we were on the homestead, Indians used to come over on the ice in the wintertime to hunt for jack rabbits and of course when we saw them coming we thought they were after our scalps. One time I was coming home from school, the little ones always got out a half hour early, and I saw these two Indians hunting with their guns, you know. They didn't pay any attention to me but I ran all the way home, I was so scared! They were hunting jack rabbits.

You asked me if I had ever met any Indian Chiefs. When I was 17 ½ years old I had the honor of meeting Chief John Grass, he was a full blooded Sioux Indian and a very fine man. My father thought a lot of him. One day I remember my father came home and he was just laughing, oh, he thought it was the biggest joke there ever was. He had an offer for me. Old Chief John Grass wanted to trade him a bunch of Indian ponies for me the one with the dark hair. My sister was blonde and my hair was dark. He wanted me to marry his son, Albert Grass. So anyway, Albert Grass at that time was over in the war, the first world war, and he lost his life in the war, so that never came about. I missed my chance of being an Indian princess, didn't I? Just think, you could have been the daughter of a princess.

Hazel Winter, 75, Barnes County

An Urban Norwegian in the Red River Valley: 1883-1885

Gunnar Høst is one of the small number of Norwegian immigrants from an urban background. Born in Moss in 1861, he was educated in Bergen. His father, Ude Jacob Høst, was the district doctor in Kinn from 1863 to 1880. Around 1872 Høst went to Bergen to attend school, and he lived for eleven years with the family of Claus Hansen. Claus Hansen worked in a bank, and the Hansens were an upper-class, educated family of scientists, businessmen, and doctors. Among the children of Claus Hansen were Gerhard H. Armauer Hansen, the doctor who discovered the leprosy bacillus, and Klaus Hanssen [sic], a pioneer in the battle against tuberculosis. Gunnar Høst was brought up as a son of this household, and he regarded the whole family with great affection. After attending business school, he worked for a short time as an assistant at Norges Bank, where Claus Hansen also worked. However, he wanted to seek his fortune in America, and left the Hansen home in Draeggen (a residential area of Bergen) in May 1883, traveling to America with two friends, Birger and Henrik Smitt. Birger later became a journalist and Henrik entered business.

Gunnar Høst writes from the perspective of a well-educated, urbanized person. Spanning the years 1883-1899, the twenty-four letters in this series were all written to Høst's foster sister, Agnes Hansen, or to Malla, another member of the Hansen household. They tell of his arrival in the United States, how he started out as a farmer and then moved into business. He owned a store in McIntosh, Minnesota, and later worked for a firm in Grand Forks, North Dakota. Gunnar Høst married Ida Olson in 1892. She was born in 1868 in Decorah, Iowa, but grew up in Minnesota. The Høsts had four sons, Sidney, Archie, Victor, and Edgar, and one daughter, Agnes. After living in McIntosh and Grand Forks for some years, they moved to Leeds, North Dakota, in 1901. There Gunnar Høst established a store, which he and his sons ran for fifty-eight years. He served as mayor of Leeds from 1923 to 1927. Mrs. Høst died in 1925, and Gunnar Høst died in 1943.

As we would expect from an educated person, Gunnar Høst writes well, using standard spelling and grammar. He gives detailed descriptions of his activities and discusses his feelings openly. His affection for his foster family comes through clearly in these charming letters.

At the time of this first letter, Gunnar Høst and his friends Henrik and Birger Smitt have begun farming in the Red River Valley, near Stephen, Minnesota. The reference to "my friend Bjørnson" in this letter is to Bjørnstjerne Bjørnson, a Norwegian author who held many controversial social and political beliefs. Bjørnson traveled and lectured in the United States in 1880-81. Gunnar jokes about his friend Nikolay Martens and his bread. The Martens family owned a bakery in Bergen.

Stephen, Minnesota
July 18, 1883

Dear Agnes,

Many thanks for your dear letter. I was especially surprised to receive an answer so quickly. It was nice to hear about the wedding and about all of you. From my letter to Malla you can see that we have taken land up here and have begun working for ourselves. We are trying to get our land under cultivation before spring so that we can plant wheat, which is the only crop found in great quantities up here. We can't have much livestock because it is too cold in the winter, but we are thinking about getting land near the river, where it is wooded, then we would move down there in the winter when the prairie is so bitterly cold. If we do that, we might go in for raising cattle. The prairie landscape is not in the least poetic, not a bush nor a rise in the ground can be seen, though it is not so bad here where we are, for we can see the woods in the distance and that immediately makes the landscape livelier. We often sit outside in front of the house in the evening when the sun goes down behind the edge of the woods, and then it is really rather beautiful. One eventually gets used to it. I would never have believed that I would find beauty in a landscape without water, mountains, or forests nearby.

Our work day lasts from six o'clock in the morning until six o'clock in the evening. The most pleasant time is before and after supper. Before supper we sit outside and follow the sun—yes, ask it to greet our dear ones in the east, knowing it will awaken you in a few hours—and after supper we sit and chat over a cup of tea and a pipeful of poor tobacco. You ought to see me do the housework, cooking, baking, and washing. Tell Father that it is coming in handy now, that I learned a thing or two during the time I lived in Dræggen. . . . When you write to Salma you must tell her that I am a master of potato cakes, though mine are still not quite up to the quality of hers. We frequently dine on duck, as there are lots of them around here. We most often serve them boiled, but on Sundays I brown them á la Salma in sour cream and butter. . . and ours taste *almost* as good as hers. We don't prepare anything other than duck, since we can't afford to buy beef or pork, which is terribly expensive. I am best at baking bread. I am certain that my friend Nicolay Martens wouldn't believe that I have baked our bread this week, it is really splendid. As you know, we live rather frugally, flour and water in different combinations, such as porridge, pancakes, lefse, etc. Anyway, now we have gotten two chickens, which lay every day, so our pancakes are superb. You should have seen us one day, we were so fancy that we treated ourselves to soft-boiled eggs for supper, how our faces beamed. No, this is getting to be completely a "food letter". . . so enough of that for this time. . . .

I have not written before and told that when we got to New York we had to go into quarantine. You see, there was smallpox on board, luckily the first-class passengers got only one day, where the third-class had to stay eight days. The reason I haven't mentioned this before is that I did not want Mrs. Smith to know about it in case she would worry. . . . But now two months have passed, so I am sure we can be regarded as not contagious, which, of course, we were declared to be in New York. The day we stayed in New York was very moving, everyone thought it was sad that we should be parted and so we decided to stay in the same hotel, the whole first class. You can imagine how comical we looked when we came walking through the streets of New York all in a line. There were many good wishes and promises to write when we departed from our friends after having spent a pleasant evening together in New York. On the railway journey we met many friendly people, but since I had gotten instructions from my American friends to take care, I always answered coldly and shortly, so H. and B. often thought I was a bit curt.

Yes, at this time of year it is wonderful in Bergen, we really miss the long, light evenings. Here it gets dark as soon as the sun goes down. It is really comical to see America close up and compare it with the reports we heard. I especially have to laugh when I think of my friend Bjørnson's description of the intelligence, freedom, and equality of the Norwegian-Americans. Class differences are absolutely as strong here as in Norway. "Big farmers" keep together and naturally exclude small farmers and workers—and it isn't the core of the Norwegian nation that emigrates, it is cotters and small farmers who have difficulty making it at home who seek a new home in America. Those I have met have been decent and good people, but quite astonishingly low class. . . . The worst, I think, is that they do not have enough concern for their children's education, so that they don't have the opportunity to rise much above their parents' station in life.

I was pleased to hear how you appreciated my letter from the Atlantic Ocean. I thought I had to write to you before I wrote to anyone else. You were sister, mother, and friend to me and followed me with interest during the eleven years I lived with your family. . . . I lived a blessedly happy adolescence in Bergen, and yet I am happy to be in America with a bit of my youth left. It is good to get out a little, and especially I believe it is good for young people to come to America. If I had come over here alone, I believe I would never have been happy, but since there are three of us, then we all get along remarkably well, and yet I always work with the thought in the back of my mind that someday I will return to Norway again. . . .

You must promise me that my letters won't get outside of Dræggen, I cannot bear to have my letters read by everyone. They are for you and Father and Malla. Of course, I don't object if you tell what you want of what is in my letters. . . .

A thousand greetings to Father. . . and all the others in the family. . . as well as any others who ask about me. Live well, dear Agnes, and greetings from your loving,

Gunnar Høst

Christiania (also spelled Kristiania) was the capital of Norway. The name Oslo was restored in 1925. This letter was also sent from the farm near Stephen, Minnesota, in the Red River Valley.

September 11, 1883

Dear Agnes,

I received your postcard from Christiania yesterday, and am glad to see that you have had a pleasant trip. . . . Here we are all still fine. We have had a beautiful summer with quite a bit to do. Now it is getting along toward fall, so we must begin to think about going in to town to seek winter employment. At any rate, we will first go out for three weeks and earn money during the fall work. I am going to work for a Frenchman and will get two dollars per day plus room and board. I believe I will begin to miss this place if I go into town this winter. I have become so accustomed to the landscape now, and will be longing for next spring when I will come up again to begin the plowing. I have high hopes of getting a really good farm up here, if I just have enough money for the first plowing, and that I hope to earn this winter. We have cut a great deal of hay this summer, and have been out working for other farmers. I am finding the work very agreeable, and have been poorly just once. It was my old stomach problem, and went away quickly. . .

Birger and I were over on the other side of the Red River in Dakota last week to look for work. We thought we would reach the town of Grafton in the evening, but the night became so dark that we had to sleep on the open prairie with our coats around us. But we slept well and fortunately suffered no ill effects from this. Henrik and Birger have both been healthy the whole time and, like me, are very happy here. When we have been gone a couple of days, we take great pleasure in our cozy little home and exclaim, "How nice it is here." As we now have high hopes for the future and good health, I can't think of anything we are lacking.

No, now I have to fix supper. You won't hear from me again until I return from the fall work. Live well and greet everyone. . . .

From yours,
Gunnar Høst

The ink in this letter has faded, and some words are completely unreadable.

October 7, 1883

Dear Malla,

Yesterday I received your dear letter of last month; it is a long time since I had heard anything from you and it was a great joy to hear that you are all well. . . As I wrote to Agnes, we went out working; we all three worked with a threshing machine in a French colony on the other side of the Red River. We got two dollars a day and free board. Now we are home for a few days before going out again later in the week. It is pretty hard work with the machine. We had to get up at four in the morning and work until seven in the evening with one hour for dinner. The food was remarkably good; you see, the machine went around from farm to farm, and as the farmers over here help each other with the threshing, the one Mrs. would outdo the other in the matter of food, which was very fortunate for us. I like the work, but there is one drawback: it is beginning to get rather cold here now. We have even had several nights with frost; but I hope we will still have some good weather in October so that we can earn enough money to be able to go down to a larger town, where we could get work through the winter.

We got a real fright when we returned from our work. We had been told that the whole prairie on this side of the Red River had burned. Of course, we had taken our best clothes up to our neighbor, but the thought that our cozy little home should be burned up was most unpleasant. Great was our joy when we caught sight of our little house and our two stacks of hay. We found all as we had left it, but to our sorrow, heard that our neighbor had lost two haystacks in the flames. As he has a family and cannot get anything off his land before next year, we decided to give him one of our stacks of hay—the fruit of many days of hard work. In return we got lots of potatoes and frequent invitations to his home. Lately our kitchen has been very poorly provided for, mostly flour and water in the form of bread, porridge, pancakes, and potato cakes.

It is strange to live so completely cut off from the civilized world. I have only seen newspapers a couple of times since I came out here, so I have no idea how things are going out in the world. So please write a bit about political conditions at home, as well as what new things in literature have come out since I left. Greet everyone and. . . write often to your faithful son and brother,

Gunnar Høst

Stephen, Minnesota
October 28, 1883

Dear Agnes,

Well, now I am about to change tools. I am sick and tired of darning socks, and will therefore take up my pen to write a few words to you and thank you for your dear letter, which I received the other day when I came home from a trip to Caledonia. You see, I have been on an errand down to Mr. Jacobsen, the man we came to last spring. You can believe I had a good time down there! Mrs. Jacobsen did everything she could for me, and I just enjoyed myself. Yesterday Henrik went down to wait at Jacobsen's for Birger and me while we get ready for our trip to Chicago. We are thinking about spending the winter in Chicago, you see, or possibly St. Paul. Later this week Birger and I are going to take a trip over to Dakota to sell some hay. Tomorrow we are going hunting as we are out of food—this time it's prairie chickens that will fall to our sure aim. They taste splendid and are about the size of regular chickens. We make soup out of them and brown them in the pan. At the same time as we go hunting, we hope to get a small pail of sour cream, so we can fry pancakes with sour cream and live like kings. . . . We have been busy lately washing and repairing our clothes, which isn't much fun.

I don't need to tell you that I get along with my two friends remarkably well. . . . Henrik is unusually practical and clever with his hands; Birger has enough energy for all three of us and is the noblest person under the sun. They both like the country life best of all, so it is with sorrow that we think of parting from our dear home on the prairie, but that must now be done.

All summer I have been collecting butterflies, and had a nice collection. Unfortunately, the other day the wind almost completely wrecked it for me. I had intended to send them over to Malla, but now I will have to owe her until next year. There are some beautiful specimens here.

This evening we are going to have porridge, as we got a pailful of sour milk from a farmer because I wrote a letter for him. Dear Agnes, you must not be angry with me because my letters are so dry and short, but there is little here to write about, as we are so isolated. Greet the family and others from me. . . .

Gunnar Høst

The end of this letter was written around the edges of the paper, and was quite faint and difficult to decipher. It was most likely written in 1884.

Grand Forks
September 24, 188[4]

Dear, dear Agnes,

I received your wonderful letter of August 25 just today, and want to thank you so much for it. You have no idea how wonderful it is to get your letters. . .

I came to Grand Forks four weeks ago as wages here were higher than close to the farm. I was lucky and got a job for $2.50 per day, but after four days of work I had to come into town as I had gotten a bad case of jaundice. The doctor forbade me to take any kind of work until I was completely cured. You can well imagine it costs a pretty penny to lie here with doctor bills and medicine to pay for. Yet one must simply accept it. It is hard to lie still now at the best time of the year, when one should be earning for the winter. If everything had gone well, I would have earned about $100, and that would have been enough for the winter. The other day I felt strong enough to go out working again, and then they put down the wages and all the workers went on strike. Now it's like this here in America, you have to go along if you don't want to risk getting shot, and therefore I also had to strike after having worked only a few days, even though I would rather have worked for $2.25, as it always difficult to get a good place again, and besides, it's a waste of time. I am still not well, but will just have to keep at it. The jaundice has worn me out, and in addition, I have a complaint in my chest. I don't know what it is, but it bothers me when I am working. I hope it will soon get better, otherwise it looks dark for me.

I have not decided what I will do this winter, but most likely I will stay up here. I may be able to work for a doctor. He has a farm six miles from town and wants a man out there to take care of the cattle and the farm. In the spring I am planning to go down to St. Paul where I can get work for the government for $100 per month, as well as everything, including clothing, free. It's true that I have to bind myself to stay for three years, but there's no help for it. The work is easy and only eight hours a day. Most likely I will be sent west to the Pacific Coast. When the three years are up, I intend to come to Europe again, to look around. If I could find something at home in Norway, I don't think I would come back here again, but much can change in that time. Don't talk to anyone about this, as it is still undecided if anything will come of it. . . .

As you probably have heard, Henrik has gone to Watson, Minnesota, where he has gotten a position in a business. . . . Thus our partnership has been broken. We are having wonderful weather here, and the harvest work is going well, but it is hard to work from five in the morning till nine in the evening and then lie in a shed without bedclothes. In four weeks I think most of it will be finished and then we will creep into our dens for the winter; that's what's so bad up here, for seven months of the year everyone goes completely into hibernation.

Live well now, my dear Agnes, and heartiest greetings to all in the family. . . .

Your always grateful brother,
Gunnar Høst

In this letter, Høst begs Agnes to send him the latest book by the great Norwegian playwright, Henrik Ibsen. The Wild Duck *came out in 1884.*

Grand Forks
November 13, 1884

My dear Agnes,

Many thanks for both letters, which I have now received. You have no idea how pleasant it is for me to get your letters. I become like another person just by receiving them. The letters from home are always more depressing, as I am not spared from hearing about my father's rather poor economic circumstances. You can imagine how that discourages me, as I am under great obligation to him, yet am still not in a condition to do anything. Every time I get a letter from home, I open it with fear, and as often as not I am in rather low spirits when I have read it.

As you probably already know, Birger is no longer with me. He went down to St. Paul three weeks ago. . . . Since then I have been out working and have done quite well. We have gotten snow now, and intense cold, so all work is stopped. I took a trip down to the Norwegian settlement near Mr. Jacobsen to see if I could get a job as a schoolteacher down there this winter, and was lucky enough to be hired. First I must travel up to our land to pick up my things and then I will go down to begin the school. It won't be any great wages, but one must be glad to have something. As far as comfort is concerned, that too will be sparse, as the farmers down there live anything but comfortably, in my opinion. Still, I am quite happy, and will just have to take pleasure in my books and the children this winter. Most of the time I will teach in Norwegian, giving instruction in religion as well as arithmetic and writing. . . .

I am now free of my jaundice. Thank goodness it went away quickly, and since then I have been well. The other day, when I stopped at a hotel in a little town by the name of Fisher Landing, I was served with—potted meat—made just like you used to make it. Then I couldn't help thinking about the pleasant fall and winter evenings in Dræggen, and it cannot be denied that tears threatened to flow. I often become quite melancholy when I remember my life with you, many times I wish I could make myself invisible and fly home to you even if only for a few hours. And yet even though I often miss good care and pleasant family life, I have never regretted that I came to America. I feel so strongly that I could never become a *man* at home.

If you should see Miss Kjeding, you must tell her the following. I came into the hotel here after having been out in the country working for four weeks. As soon as I came into the kitchen (I am always good friends with the kitchen personnel) I was told that a terribly fine lawyer from Minneapolis had come in. I was very curious to see this fine gentleman, and met him then at the supper table. He condescended to speak to me, and it turned out that the fine lawyer was a Mr. Boyesen, one of Gina Kjeding's good friends from Chicago. He had been watching me and wondering what kind of a fellow I was, of the "better class" and yet wearing working clothes. Since then, Mr. Boyesen and I have become good comrades. He dupes people with his fine exterior, while I stick to my working costume, though in the evening we discard our masks. . . .

I am ashamed of what I am going to serve up next, but the matter is this: I have read in the newspapers that Ibsen is coming out with a new book in time for Christmas, and my dearest wish is that you and Malla will send it to me. I know it is terrible to beg, but I am just about crazy when it comes to good Norwegian literature. Ask her please not to be angry with me—ask her to imagine herself exiled among people without a trace of interest in anything other than their farms and then to understand my joy over receiving a little intellectual nourishment! Finally, my own dear Agnes, many thanks for the paper and envelopes. How wonderful it is for me to receive a sign of your affection. Greet everyone in the family. . . . from your devoted,

Gunnar Høst

Grand Forks
November 25, 1884

My dear Agnes,

If you are to get a Christmas letter, it is time I took my pen in hand. Since my last letter I have been up to my land and gotten everything there in order for winter. I had a rather hard trip, as it was terribly cold. This afternoon I will go down to take over my post as schoolteacher, which I told you about in my previous letter. But I didn't tell you that there was a student who also wanted the position that I got. He was going to teach in another district a few miles from there, and he wanted to have both schools through the winter. However, the farmers thought it best that we divide it, and so, as I said, I got the post. In the meantime, I had not met the student and didn't even know his name, but I was a bit anxious that he would eliminate me from the competition, for as a student he would naturally have far greater knowledge than I. When I got to Crookston, I found out from an acquaintance that this student was none other than—can you guess?—Fredrik Mohn—Uncle!!! I was told that he had been seen that day in Crookston. I went downtown immediately to look for him, but unfortunately he had already left. However, I hope to meet him either this evening or tomorrow in Fisher Landing or Sandhill. Then I can relay greetings to him from his parents and brothers and sisters. Isn't it remarkable how people can meet up with one another. You can imagine how I am looking forward to seeing him. . . .

Well, now Christmas is fast approaching, and then, you know, my thoughts fly over the Atlantic Ocean. I am wondering where you will all be on Christmas Eve. Klaus has first priority, of course, but on the other hand, Sofie would surely like to have you together with her family. . . . I have already been invited out for Christmas Eve, to Mr. Jacobsen, as well as to Grand Forks.

My dear Agnes, I send you my dearest greetings and wishes for a blessed Christmas holiday for all of you, and do not forget me when you sit in your pleasant family party.

Your gratefully obliged,
Gunnar Høst

Caledonia, Dakota
December 10, 1884

My dear Agnes,

Well, now I really must write to you and tell you about my meeting with Fredrik Mohn. You can just imagine how surprised he was to meet someone from Bergen who came with greetings from home, even if they were rather old. I was together with him for two days in Fisher Landing, and we had a really good time. He thought that everything I told him about Bergen was so pleasant, and I am never happier than when I can talk about Bergen and my dear ones there. You can imagine how surprised I was to find Fredrik as a totally healthy, noble, and vigorous young person. The fact of the matter is, I thought that by studying theology, if I may say so, he would have become intellectually crippled, but quite the opposite, he is imbued with a true, inner, healthy Christianity and is an outstanding person. I have not heard him preach yet, but from others who have heard him I know that his sermons are supposed to be very good. What he most lacks is self-confidence; he is almost too modest. In the old country modesty was supposed to be the greatest virtue a youth could possess, but here in America it doesn't do to be too modest.

I went up and visited him the other day. You see, his school is about five miles from here—and then I brought some photographs along, and you can imagine what a pleasure that was for him, as he recognized old familiar faces. I had planned to make a short visit, but ended up staying the whole day. I think it is so pleasant to have him nearby, as I have become very fond of him.

This Christmas we are going to have a Christmas tree for our pupils; we two are going to go together to make a pleasant celebration for the children. Yes now I have started as a schoolteacher, and I think I will like it. I don't know yet if I will be any good at it, but I hope everything will go well. I have been offered a teaching post at twenty dollars a month starting this spring at a place four miles from here—I don't know yet if I will take it.

Christmas Eve I will be with Mr. Jacobsen and I have also been invited out on the other days of Christmas. The young people are going to have a sleigh ride and I have been asked to go along. This will no doubt be the last letter you will receive from me in the old year, so . . . thank you for each loving thought and for all your dear letters in the past year . . . Greet everyone. . .

From your devoted,
Gunnar Høst

Høst has received a copy of Alexander Kielland's novel Fortuna, *which came out in 1884. Kielland's novels were sharply critical of contemporary Norwegian society, and in* Fortuna *he concentrated his attack on hypocrisy in business. Høst refers in this letter to "Christmas Fools." The Christmas Fools, or Mummers, traditionally dressed up in masks and went in groups from house to house, demanding treats at Christmastime, rather similar to the American Halloween tradition.*

Caledonia, Dakota
January 14, 1885

My dear Agnes,

Well, now, it was really kind of you to write to me again. I had been waiting for a letter all week, and it finally came on Saturday, just as I was about to go north to hear Uncle preach on Sunday. I also received *Fortuna* in the same post. . . . I threw myself over *Fortuna* at once, it is very well written, and the descriptions surely true, but I found it all terribly depressing. When I was at home, such books didn't make as strong an impression on me as they do now. I don't know the reason, but I believe that I now need a different kind of literature to feel satisfied. Perhaps it comes from my changed life circumstances, perhaps I just don't understand contemporary literature, as I have already spent too much time in the wilderness.

I have had a most pleasant Christmas with dancing and other fun, yes, I have even been out Christmas Fooling. What do you think of the worthy schoolmaster as a Christmas Fool? Fredrik Mohn and I had a Christmas Tree for the children last Saturday. That was also very pleasant. I like being a schoolteacher, and believe that I am well suited to the profession. The other day, one of the most respected older men in the settlement came and asked if I was the one who was so unusually good at teaching the children reading and arithmetic. You see, my reputation is growing. Well, if only I can succeed in getting something into the heads of these children, which is truly not so easy, as instruction on the whole is quite neglected for most of them. Their dear parents believe that a schoolteacher can, in the course of one or two months, teach them all the wisdom of the world. I work with the children five or six hours a day. My free time I spend studying, together with a little outdoor work and sport, namely rabbit hunting. The people I am living with are unusually pleasant, so I am getting along fine. Mail days are a real holiday for me, and I usually get quite a few letters, but my greatest joy is when I hear from my dear old home in Bergen. However, most of my mail comes from my friends here in this country. . . .

"Uncle" speaks enthusiastically of Malla; she is his ideal. On the whole, he is happiest when I go up to him and stay overnight. We disagree on a number of issues, so we have a lively discussion from the time I come until I leave. In the morning I am awakened by him as he reels of all the arguments he has thought up during the night, and so we start up again. He very much wants to have me as his sexton during church services, but I don't want to. What do you think of that, Uncle as minister and me as sexton? Ask Salma what she thinks of Høst as a sexton! I had better quit here for this evening. I am supposed to play with the children, and they are getting impatient. . . . Greet. . . . everyone who asks about me.

Your grateful,
Gunnar Høst

Caledonia, Dakota
February 6, 1885

Dear Agnes,

Just a moment ago I received your dear letter. . . and it was a great joy to hear from you! . . .

I am now working hard trying to get some knowledge into the heads of my pupils, and I believe that it is going rather well. They all like me so much, that sometimes it is almost too much of a good thing. If I have a free hour, they compete for places on my lap, around my neck, and on my shoulders, and I must tell stories. Unfortunately, I can't remember so many fairy tales, so I often have to make them up myself, and everything appeals to my grateful listeners. The little girls vie to stand high in the favors of the schoolmaster. . . . Yesterday we had a lovely sleigh ride with my pupils in crisp, cold moonlight, which you can well imagine was lots of fun.

Next week I am going to a wedding, and for that occasion I have written two songs, they were created in ten minutes, so you see it doesn't take long to make a "verse," but you can imagine how they turned out. Nevertheless, they are the best ones people around here have heard! So the poet has every reason to be satisfied.

I am very busy teaching, chopping wood, sawing, as well as taking care of the animals, so have little time to read anything for myself. . . . Is Miss Kjeding coming back to America? If she does come, there are a few small things I would very much like to have her bring over: wool jackets and underclothes. What you get over here is such poor quality that it can scarcely tolerate washing. I will send money if she is willing to bring them over with her. . . .

Live well, dear Agnes, and greet everyone. How is Salma? I think of her almost every time I eat a meal. With loving greetings,

Gunnar Høst

Caledonia, Dakota
March 24, 1885

Dear Agnes,

My last two sheets of paper must be divided between you and my parents in Moss. Yesterday I got your letter. . . and I read it over and over again and seemed to see you and the others so vividly before me. I have also had a letter from home, and it was the nicest letter I have had from them since I came to America. They would very much like to have me back again, but as you yourself can understand, it isn't possible for me to go back to uncertainty;

it would be different if I had a position to return to. If I can earn enough this summer and sell my land profitably, then I might take a trip home, either in the winter or next spring, to fulfill my parents' wishes. As for myself, I both want and don't want to come home. If I were to live in Norway, I could not imagine any other city than my beloved Bergen, and times are so bad there now that there is no future for a man without capital, though it would be wonderful to come home and see all my dear ones. But there is no place for these thoughts, they are too disturbing.

I quit teaching several weeks ago, and have now begun working. . . . I am sawing and cutting from morning to night. It is quite strenuous, but pays well. When sledding conditions break up I may possibly begin teaching school again if nothing more advantageous comes along. I talked with Fredrik Mohn the other day and he . . . is holding an evening school three miles from here. He was a bit downhearted yesterday when I talked to him. "You have made a hit here," he said to me. He has not got the ability to get along with people that I have. . . . Poor Fred, he isn't suited for the practical life, and is having a hard time fighting his way through to becoming a minister. I like him so well, and try to "push" him a little, if only it would help. . . .

Greet all friends and loved ones from your devoted,

Gunnar Høst

Gunnar Høst dreams of eating such Norwegian delicacies as ptarmigan (a wild bird, similar to a grouse) and cloudberries (a delicious arctic berry).

Caledonia, Dakota
April 25, 1885

Dear Agnes,

Many thanks for your dear letter, with the news that you have sent three sets of wool clothing with Miss Kjeding. I will send you the money as soon as I receive the package. It was more than kind of you, dear Agnes, to go to this bother for me. They will no doubt be expensive, but it is worth it for the durability.

As sledding conditions broke up some time ago, I had to look around for something to do this summer. At first I thought of farm work, but wages are so low, I decided to rent a house out on the Dakota prairie and advertise a school. So I came out here eight days ago, well equipped by all my farmer friends with all kinds of good foodstuffs, such as pork, bread, potatoes, eggs, etc. I rule the roost all alone in a house with six rooms, have nailed together a school table and bench, which is all the furniture I have. I started school on Monday and have fourteen pupils, which is really good at this time of the year, as parents have use for the children at home during planting. I predict that I will get more than I can handle once that is finished, though there are not so many families here with big children. However, they come from a long ways away to seek *the great master!!*

My hands have been terribly sore from cutting wood, and now I have a badly swollen finger. I took my departure from Fredrik Mohn two weeks ago when he went to Crookston. What he is going to do there I do not know. He was rather shocked that I was cutting wood. He, the refined Fredrik, almost threw up his hands the first time I presented myself to him in my working costume.

I hope to be able to send you a proper letter soon, now my finger is bothering me too much. . . . Greet everyone at home and in the family. The other night I dreamt that Salma was standing there crying because Høst hadn't been home on Sunday and we had ptarmigan and cloudberries (yum! yum!). I especially think of Salma now that I am taking care of myself. I wonder if I will ever again eat food prepared by her. To me, hers is the best of all food. Greet her so much from Høst. My greetings to. . . all the others, but first and foremost to you, dear Agnes, from your brother,

Gunnar

Grand Forks, Dakota
August 17, 1885

Dear Agnes,

Now it is a long time since you have heard anything from me. As you probably know, I have been working. . . at the county treasurer's office. When that was over I became a juryman and now tomorrow shall begin as a clerk in a large dry goods and grocery store. I expect that I will live with my employer, who has an unusually pleasant American family. I have been with this family quite a bit, and together with the daughters and a son I sing in a quartet in church. Yesterday I sang two solos, which earned me lots of praise. Americans are not fussy when it comes to music.

I have finally gotten the underclothes from Miss Kjeding, and they are absolutely wonderful. I will send the money as soon as people get their wheat threshed. You see, money is tight here right at this time of year, but it looks like it is going to be a good year, and then people will have cash. In a few days I will send you a photograph. I have shaved off my beard so I won't frighten you, but as you see, I have gotten old in these two years. I still have my farm, and will keep it until times get better. It looks like Grand Forks will be my place of residence in the future. It is a bit sad to give up my outdoor life, but on the other hand, one must think of the future, and for me it is more certain in town than in the country.

Live well now, dear Agnes, and many greetings to all my dear ones in Bergen. . .

From your,
Gunnar Høst

This letter is the first of many on letterhead from the Gotzian & Christoffersen Dry Goods Store. From the opening paragraph of this letter, we can infer that Gunnar Høst has had a trip to Norway. In a later letter, he mentions having visited Bergen in the spring of 1887.

Grand Forks, Dakota
December 2, 1887

Dear, dear Agnes,

Christmas will not go by before you hear from me. It is wrong, I know it is wrong, that I haven't written before, and I have no excuse other than that I have been terribly busy and that I have been waiting and waiting to hear from you. I don't know if you are aware that I am no longer working at the old place, for Mr. Griffiths. I started working for a new firm when I got back. I like it, but have terribly much to do. I had an absolutely delightful journey over, wonderful weather and pleasant company. I met up with three ladies from Christiania, who were more than happy to place themselves under my protection. I stayed in Chicago for eight days and had a most pleasant time with my old friends and acquaintances. . . .

We have had unusually nice weather this fall, still no snow though rather cold. Now you must think about me a little this Christmas, Agnes, and write to me. I miss your letters. Best wishes for a Merry Christmas and a blessed New Year to you and all in the house. . . . Hearty greetings from your devoted.

Gunnar Høst

This letter consists of loose pages, which may not all belong together. The beginning is missing. Høst mentions celebrating the seventeenth of May, which is Norway's Constitution Day.

Grand Forks, Dakota
1888

Several times I have been offered positions in other firms, but I don't think it looks good for a young man to change places too often. Besides, I have good prospects of getting into a business here, and then my future will be assured. I could have come into this firm some time ago, if I had put in some money—but it is terribly expensive to borrow money out here, so I decided to work for a monthly salary until fall in any case. . . . [The top line of the next page is missing.] to be taken into the firm without very much capital in the fall. Thus one builds castles in the air. If they stand, all is well, and if they fall down—well, you can just build new ones. That is what is so delightful about America, one has such rich opportunities for building castles in the air—here nothing stands in the way.

Well, you can believe that I have got a nice situation now—when it comes to food, that is. A widow here of a Dr. Monsen. . . . [The top of the page is missing.] As you can imagine this is very different from eating at a hotel. It is Mrs. Monsen's greatest joy to see us eating well and she seems to have devoted her life to preparing good food. The others who eat at her place are Boyesen the lawyer, Berg the pharmacist, Lohrbauer from Christiania, Mr. Sanness, and me. Mrs. Monsen reminds me a little bit of you when you smiled in great satisfaction every Tuesday as you watched me empty my plate of rice porridge.

[The top of the page is missing.] I have been waiting and waiting for the life and action that comes with spring. However, my expectations have not yet been fulfilled. Only in the last few days have we had any warmth, and with that a little bit of green on the trees. You cannot imagine how dreadful this winter has been, cold and long, though I really have nothing to complain about as I have my work indoors where I certainly do not spare the wood and coal. . . [pages missing].

Salma doesn't need to worry about me, I am getting plenty of good food. The 17th of May our Choral Society was up in a neighboring town touring from five in the morning to six in the evening, when we returned to sing in the festivities here. As a result of the singing and shouting I am so hoarse I can scarcely talk, though I expect it will soon pass. It looks like we will have good times this summer and fall and that will be wonderful, doubly wonderful if I could set up in business for myself.

Everything would be wonderful out here if we only had a little more feminine company. I hear about so many young woman who go out in the world to try their luck. It astonishes me that none of them think of Grand Forks, where we have a very respectable collection of young Norwegian gentlemen.

My best to you, dear Agnes. Greetings to all. . . and write back soon to your devoted,

Gunnar Høst

Grand Forks, North Dakota
February 22, 1890

Dear Agnes,

Now it is really a long time since you heard anything directly from me. . . . I recently had a letter from Malla—Did I say letter?—I meant a telegram. Yes, she is incredible, she can tell more in a six-page letter than most people can manage in twelve pages, but then sometimes you have to guess a little at the meaning. As you have probably heard from my family, I am very well and contented in all things. I had a letter from my mother the other day, in which she maintains that I could join in with Jacob in Christiania if I could contribute 10,000 crowns and could then work in the office. But I am afraid I have been in America too long to be satisfied to work at home. I don't know very much about the. . . business either. I hear reports from home that it is going splendidly, but one cannot always trust that kind of information.

I have worked up to a good position here and have a pleasant living situation. Three. . . gentlemen, Lohrbauer from Christiania, Boyesen from Lillehammer, and I live with a Norwegian doctor's widow, a Mrs. Birmann from Christiania, and she really knows how to make things pleasant for us. At this time of year we finish a little earlier in the evening, and then it is extremely nice to come home to a pleasant house. We have had an unusually mild winter this year, hardly cold at all. Greet everyone I know. . . from your always devoted,

Gunnar Høst

Gunnar Høst now has his own letterhead, which reads "Gunnar Høst, Dealer in General Merchandise." In this letter, he uses the English word business *but encloses it in quotation marks. There are no letters in this collection between 1890 and 1892, so some of Høst's letters are probably missing, including those telling of his marriage in October 1892 and of starting his own business in McIntosh, Minnesota.*

McIntosh, Minnesota
December 5, 1892

Dear Agnes,

Many thanks for your exceptionally pleasant letter. Ida was overjoyed to receive it. The first thing she asked, when I said I had a letter from Norway, was whether it was from "that one who looks so nice." You see, she had just remarked a few days earlier on your photograph. I am very happy in my home, and Ida is an unusually excellent housewife. My business is going very well, and yet it has caused me many agonizing moments. There was a man who was going to invest $2,000 in my business, but when it came time to pay, he couldn't get the money—he had sold his farm but couldn't get the cash. However, I had organized my business in the expectation that I would get the $2,000, and there I stood! My business has, as I said, gone very well, but it is almost too much to manage. It is nearly impossible to borrow money out here, as it is so expensive. A loan of $1,000 would cost me $100 in commission plus twelve percent in interest, and that is backbreaking. If one could only borrow money at a reasonable interest rate! But what am I thinking of, writing about business to Agga? It is just that I am so concerned about all of this that I have no peace, either day or night.

Well, now it is getting close to Christmas, and this was supposed to be a Christmas greeting to you. I sincerely hope that this letter finds you and all my dear ones healthy and hearty. Best wishes from both me and Ida for a very Merry Christmas. Hearty greetings,

from Ida and Gunnar

In describing his home, Høst uses the English words for parlor *and* bedroom, *but the Norwegian word for kitchen. As least one page is missing from this letter, and one page included here may actually belong to another letter.*

McIntosh, Minnesota
October 21, 1893

Dear Agnes and Malla,

No, now I must write to you and tell a little about how we are. I have been waiting and waiting for better times—it is so difficult to write when something is weighing one down—but in vain. It just gets worse and worse. The only ray of light is in my home. And now I must let you know that I have become a papa; on September 22 Ida gave birth to a large boy (eleven pounds). It went well and both she and the boy are fine. I would rather have had a little girl—yes, I had even picked a name for her. She was to be named after you, Agnes, and my mother: Agnes Lorentze. What kind of name we will give the boy I don't know yet. Norwegian boys' names don't sound good in English. . . .

Oh, I guess I didn't tell you that I have built a little house. It went up in nine days, and consists of three rooms—parlor, bedroom, and kitchen. That's right, Agnes, will you please send me your recipe for potted meat, as well as for Salma's pork pancakes and rice cakes? I am longing for some Norwegian homecooked food. As you probably know, we are living in a period of monetary crisis, a crisis that I don't see any end to. God knows how it will go—one must do the best one can and let come what may.

Ida's sister was here for a time while Ida was sick, but now we are alone. It wouldn't be possible in Norway to take care of one's baby, do the wash and all the housework, but I couldn't convince my wife to have a girl to help for anything in the world. Here one is respected and admired for the work one does, be it physical or intellectual labor. Let me hear from you soon. The large picture is a great joy to me. I chat with you all now and then. Greet everyone from yours,

Gunnar Høst

In describing his baby son, Høst writes the word bright *in English, but puts it in quotation marks.*

McIntosh, Minnesota
December 4, 1893

Dear Agnes,

Many thanks for your most welcome letter. It is so nice to get your letters telling about everything. Today we are coming for a Christmas visit—my little son and I. It is not such a good picture of the little boy because he wouldn't sit still—and the light was too strong for his eyes—so he doesn't look as "bright" as he is. But for a not quite two-month-old baby it isn't so bad. I thought you would like to see how we looked, but don't give it out for public criticism. You can get another one when he is a little bigger. Unfortunately I have no photograph of Ida. She should have had one taken with the little boy, but to tell the truth, she didn't have a decent dress to use, since all her old dresses have become too small and she hasn't had time to sew any new ones. I can't get her to have any made, she wants to do it herself.

Many thanks for the recipes, a little bite of "Dræggen" food will surely taste good. For Christmas we are going down to Grandmother's house, and Ida and the boy will stay there all of January for her to rest and get her strength back. Here we are still fighting bad times. They are trying to get people to believe that better times are in sight, but it doesn't look good. God knows how it will end. Our little boy still has no name. He will be baptized at Christmas time down at Grandmother's house. He is chubby and fat and seems to be doing well. He is already ten weeks old and weighs eighteen pounds.

We are now in the midst of winter with snow and biting cold. But life goes on as long as one has a reasonably warm house.

Christmas is approaching fast—if only we could peek in on you—but that is not possible, so we will just send our best wishes by letter. . . . You won't forget to greet Malla and Salma, as well as anyone who cares to get a greeting from me.

Your brother,
Gunnar Høst with wife and son

McIntosh, Minnesota
February 2, 1894

Dear Agnes,

The other day I sent a picture of Ida, so that you can have some idea of what she looks like. It isn't a good picture, as she has a melancholy expression that is not natural for her. . . . I have never liked this picture, but thought just the same that now I had to send it or else you would think I was ashamed to show off my wife. She is no great beauty, but nevertheless prettier than the photograph shows, tall, with an attractive figure.

We spent Christmas down with Ida's parents, where we had a very pleasant time. The intention was that Ida would stay there through the month of January, but after I had left, she began to get homesick and came back on January 4th. We baptized the boy while were down there, and named him Sidney Clifford. You probably think it strange that we didn't give the boy a Norwegian name, but the reason is that he will no doubt live his entire life in America, and so he will get a name that fits.

Many thanks to Malla for the Christmas greeting, it was wonderful. . . I began to be quite homesick for the mountains and the ocean. I always used to think it was an affectation when anyone talked about longing for the sea, but one who has grown up near the sea and then comes out to America's plains certainly will feel it—and I have felt it, too.

Our little Sidney is getting along very well—growing big and strong and he is very good.

This year we have had an unusually mild winter, which is a blessing in these bad times. You can have no idea how terrible it is in America right now. Millions of people without shelter and food. What the end will be is a common question, and no one can answer it, no one can give hope and comfort to the hungry masses. No, let me stop before I get too far into this sad theme.

Tell Salma that the potted meat was excellent and that I am now eating pork pancakes and rice cakes every week.

Greet everyone in the family and others who remember me. Your brother,

Gunnar Høst

The "Sigurd" mentioned here is probably Sigurd Scott Hansen, a member of Fridtjof Nansen's expedition that drifted in the polar ice on board the ship Fram *from 1893 to 1896. The expedition members were brought safely to Vardø, in far northern Norway, in 1896.*

Grand Forks, North Dakota
September 1, 1896

Dear Agnes, Malla, and all,

Well, it is a long time now since you have heard anything from me. First and foremost I must report that we have a little daughter, born on the 9th of June. Her name is Agnes Lorentze, after you, Agnes, and my mother. So many wonderful memories are attached to that name, that I thought I had to have it in my own family. We are all very well. Ida and the children are visiting her parents right now and having lots of fun.

Times are very bad in America, though I cannot complain as I have a good livelihood. I was surprised by a visit from Birger Smith a week ago. He came up here on a survey. He might settle down here as an illustrator for a newspaper. You must all be very joyous over Sigurd's happy homecoming. People here thought I was crazy when I broke out in cheers upon reading about Nansen's and Sigurd's arrival in Vardø.

Now I must quit. Greet everyone, don't forget Salma, and let me hear from you.

Sincerely yours,
Gunnar Høst

Grand Forks, North Dakota
May 24, 1899

Dear Agnes and Malla,

Lude has just arrived and has been down here with all the gifts. I don't know how I shall begin to thank you enough, I was completely overcome. You are certainly more than kind to think of us. Little Agnes was so proud of her beautiful Norwegian flag—and believe me, Agnes, it will always be held high in honor and remembrance of Aunt Agga. Rolfsen's reader is excellent, and I have had all I can do to tell the boys about the many wonderful pictures. It was so nice for me to get the book about Bergen, with all the wonderful, well-known faces and the biographies. I don't know how long it has been since I have had such a pleasant evening as yesterday when I read it. Many many thanks for all the wonderful books. You can't imagine how nice it is to get something from home. And then the pictures of Lorentz and Sofie's children! I have been sitting with them in front of me for many hours and never tire of looking at them. What marvelous faces and all so different. . . . You must greet Lorentz and Sofie and thank them for the gift. I cannot describe how much it means to me. Lorentz has always been my boys' ideal. I have told them about him and his ship, and that is something that appeals to a child's imagination. They always want to show off the picture of Uncle Lorentz who has the big ship.

Here we have finally gotten summer and it feels good after a long cold winter. My three little ones are outside playing all day long and enjoying life very much. They are all healthy and hearty, if not always as nice and well brought up as we could wish.

Everything is going well for me. I have a good income, that is to say, I have a salary of $25.00 per week and can earn a little extra and that is not so bad. Two years ago I bought a house with a small yard, so we live in our own home and have all reason to be satisfied. Today I am sending you the program from a concert we had a while ago, together with a description of our 17th day of May celebration. You must show the program to Tommy Hagerup so she can tell Edvard Grieg that we are trying, even out here on the outer edge of civilization, to interpret his wonderful melodies to the best of our poor abilities.

I see Pastor Hovde now and again, but our Norwegian ministers out here are so shortsighted that it is almost impossible to have anything to do with them. We have three different Lutheran congregations here in this town. They all have a monopoly on heaven, and they all state boldly that they can prove by the Bible that all the others will go to hell. Pastor H. thinks I am taking a great responsibility upon myself by sending my children to the American Sunday School instead of the Norwegian, but since my children don't understand a word of Norwegian and there is never a word of Norwegian spoken in my home, then I certainly don't understand why they should have their religious instruction in Norwegian. Of course my children will learn Norwegian, but first and foremost they are Americans and shall be brought up as American citizens. And as far as religion is concerned, I am quite certain that Our Lord understands English as well as Norwegian, even though one of our honorable pastors up here assures me as to the contrary!

. . . You must excuse me for not writing sooner and thanking you for the Christmas letter, but it is so hard for me to write, especially to answer your letters. I feel my own incompetence and am ashamed. You know I never write Norwegian except for a few words home now and then, so I get completely out of practice. I never dare to read my letters through, for then they would go right into the wastebasket. Excuse me, and take me as I am. It is all well meant.

Greet everyone within the family and the others, too. If only I could see you all again some time. Don't forget Salma. I think of her often and pray that things will go well for her in her old age. Thank Lorentz and Sofie. I don't dare to write to them, but you may be my interpreter. Hearty greetings from,

Ida, the children, and your brother,
Gunnar Høst

Grand Forks, North Dakota
October 8, 1899

Dear Agnes,

Well, what do you think of me, letting your dear letter go so long unanswered, though I hope you will excuse me when you hear that it is work that has taken up my time. I have been working from seven in the morning until ten in the evening for a long time now, and then I don't feel like writing when I get home at night. However, this evening my employer decided to quit at nine o'clock and therefore you have me here this evening. My letters can never be as interesting to you as your letters are to me, for everything you write about I absorb eagerly. Every name and every event reminds me of my own dear old Bergen.

Now I will begin as usual to write about "myself." You already know that I have an exceptionally good position. I like it very much here, and just the other day had the pleasure of hearing from my employer's own mouth the most flattering comments on my business acumen. It is always pleasant to work for someone who recognizes that one is striving to fulfill one's work obligations as conscientiously as possible, and especially when this recognition takes the form of a raise in wages. We are having extra good times here right now since wheat prices are rising. My only regret is that I have not broken up more of my farm, as it would be a gold mine if war should break out in Europe. And it looks very much like that might happen. Even so, if times remain this good, I will earn good money on my farm.

I have heard nothing from or about Fredrik Mohn. We parted as not very good friends, as he could not tolerate my criticism of him, although he asked my opinion. Likewise, he got angry when I encouraged him to give up theology, as there were a number of things that seemed to make him unsuitable as a pastor, especially in a time like this. He became very offended, even though it came after he had repeatedly asked my opinion. He was peculiar and terribly stiff. . . .

It is beginning to get quite cold here now, winter is knocking at the door. But I am well in all respects so I have nothing to fear. . . . My heartiest greetings to you and to all, don't forget my old Salma (I'm sure she remembers me), from yours,

Gunnar Høst

Six years after the last letter from Gunnar Høst available to us through this collection, we have another brief glimpse into his life in this charming letter from his sister, Louise, written while she was on a visit to the United States.

Prairie Farm, Wisconsin
March 11, 1905

Dear Agnes,

I promised myself that when I had seen Gunnar I would write and let you know a little about him as I know you are always interested. I especially wanted to do this as I know how seldom he writes. Well, he has been down here in Wisconsin on a very quick trip—from Saturday evening to Monday morning was all the time he could spend here. His journey otherwise was to St. Paul and was a business trip.

I must hasten to say that it was really a joy to see him again after eighteen years have gone by. He looks like a happy man, and as far as I can see, has a healthy outlook on life and is in all ways content. It must be said in his honor that he has managed to preserve the mark of an educated man—his appearance as well as his opinions astonished me. The fact of the matter is, it is not so easy to keep oneself up, and when one succeeds without apparent difficulty, that must be remarked upon. It was truly pleasant to see him again. Naturally, I will also go up and see his wife and children, but that will probably not be before this summer, about the time I will be turning my face toward home again. My leave of absence from the school is valid for one year, and I must say that such a year in exile is very fruitful in more than one respect. . . .

When I spend some time with Gunnar, I am looking forward, through his children, to getting some idea of the schools up there. Gunnar, together with a partner, has now bought the business that he has been running lately. He expressed his hopes for the future of the business and regarded himself as one who is coming along well. His four boys are his pride and joy. They are all musical, even his little daughter, Agnes Lorentze, who he has warned me is a self-willed little thing. I take it as a given that he is a bit of a patriarch in his own home, and that his wife waits on him more than we are used to. But he speaks of her with great affection and says he is more in love with her than ever.

She is dreading a bit having a visit from a sister-in-law from Norway, but when I have lived in America for over half a year, then I am somewhat able to understand and appreciate how the free America sees fit to bring up and develop her daughters. It would be sheer foolishness to expect to find a European lady. And anyway, such a one would be ill-suited to manage the demands made on a wife and mother here. But if I have understood Gunnar correctly, then his wife is a genuine daughter of the West, working for herself and her loved ones from morning to night.

I don't need to tell you how everything I see here interests me. The land around here is rather new, the forest has given way to farms and only in a few isolated places are there any woods left, but one is not, as in Dakota, forced to burn coal. It is very pretty here, as the land is quite hilly and rolling, but water, the sea—oh, no. One has to get along with a little lake every now and then, and you know that is nothing. . . . But Gunnar, who is used to the ugly, endlessly flat Dakota prairie, thought it was really beautiful here.

I read your memoirs this summer at Jacob's. They brought me such living memories of the time I spent with you in Bergen. . . . But people are so much more than books and possess a strength that surpasses time and place. I felt some of that now, when I found Gunnar again, so unchanged. He was fortunate in his youth to be surrounded by all of you, who encompassed him with love and care, just as if he were one of your own, and the character that this gave him has become his for life.

. . . Dear Agnes, stay well until next year, when I can come and greet you in person.

Your devoted,
Louise Høst

Two Views of the Depression

By Richard Critchfield

Those Days: An American Album by Richard Critchfield is the source for the next two selections. Originally from North Dakota, Critchfield has chronicled his family's history beginning with the marriage of his father and mother, Jim and Anne Critchfield, and ending with Anne's death in 1982. The two excerpts included here describe events from the late 1920s and the 1930s.

Jim, a farmer turned country doctor, is the person "speaking" in the first selection, "A Country Doctor Remembers the Depression." Even before he married Anne, he was known to be a drinker, and the inclination had not decreased at this time. The person referred to as "Norma" is a young woman with whom he had been having an affair. The baby sleeping on the front porch (nicknamed "Pat") is the author. A few years later, Jim's drinking problem increased until he finally died of alcohol degeneration of the brain at age 48. The author was only seven at the time, so he had to rely on letters, diaries, newspaper clippings, and interviews with family members and friends to construct his father's point of view. "Growing Up During the Depression" is based on the recollections of Richard Critchfield's older brother, Jimmy, who was a teenager at the time.

A Country Doctor Remembers the Depression

The Krause murder was just one of the grisly symptoms of breakdown as what was now becoming known as the Great Depression grew worse. The strains from bank failures, low prices, foreclosures, drought, joblessness, anxiety and despair were starting to tell. To make ends meet, I'd run for County Coroner in 1928 and since been reelected. My very first case had been an old German-born farmer out in Heimdal who'd been about to lose the land he'd cleared and settled fifty years before; he'd strangled himself by twisting a sheet and tying it to a bedpost.

It hadn't happened overnight. Seven thousand American banks, most of them rural, had failed between 1920 and 1930; seven thousand more were going to fail before it was over. The Dust Bowl was also slow in coming. Much of the Great Plains got less than twenty inches of rain in a year. Once settlers broke the sod with their plows and successive generations overgrazed and overplowed, the dust started to blow. In Fessenden, the dry years began in 1924. In 1931, the great storms, blowing away whole farms, like the Armistice Day black blizzard in Kansas, when men would vomit dirt, were still two years ahead of us. Already it was getting bad. We'd go for months without a drop of rain. When I made calls out in the country, I'd notice how some fields were slowly getting drifted over by sand. Roads, trees, sheds, fences and abandoned tools and machinery all showed signs of the steady movement of the soil. You'd see crops that had just shriveled up, cattle that were just skin and bones, abandoned farms with dried-up wells and broken windmills, and even when the sky was clear and as blue as could be, there was always a thick haze of brown dust along the horizon.

When it got real dusty in town, so that Main Street looked like a cavern in a cloud of smoke, they'd string light bulbs across and keep them going night and day so you could see where you were going. Up at our "little house on Quality Hill"—a humorous nickname just like "Poverty Flats" across the tracks, which actually wasn't much different—the dust would seep in everywhere. Anne stuffed rugs under the doors and put wet rags around the windows to catch the dust and keep it from coming in. Even so, a film of dust soon settled on all wooden surfaces. Even food tasted gritty. If a wind came up, several inches of dust might drift across our open front porch.

When it got too bad, and I had to go outdoors, I'd tie a mask over my nose and throat. Then you couldn't drive your car. When I had to make a call way out in the country, I left the Buick in our back shed and hired a team and buggy from a farmer on the edge of town, trying to get horses I could trust to keep their bearings. They'd also need dust masks of gunny sacking to tie over their noses. Sometimes out on those dirt roads, you might get swallowed up for a spell by an impenetrable black cloud. It could get dark right at noon. "That's our good topsoil blowing," everybody would say. At the height of a real dust storm, it was like a blizzard. You couldn't see a house from a barn at ten-thirty in the morning. With that hot wind howling and everything all black, it could be terrifying.

On days when the dust didn't blow and there was a bright blue sky again, Anne would put our baby on the front porch to sleep. Even when it got real cold. She'd done it with all the kids. When it started to snow and she still put his buggy out, little Billy, now seven, protested, "Mom, the baby will freeze!" "Don't worry," she told him. "I bundled him all up." When she first brought the baby home from Minneapolis, I called him Oscar, saying that he looked like a little Norwegian to me. Then I kidded Anne that he had a closer resemblance to an Irish bootlegger in town and I started calling him Pat. I was just teasing her, but though we baptized him Richard, within the family the nickname stuck. He was a skinny little rascal with big ears.

Once Anne was back home, I told myself I'd break off with Norma for everybody's sake. There were tears and recriminations, also threats. She said she couldn't live without me. She said she'd turn on the gas. And maybe five minutes later she'd be asking if the seams on her stockings were straight. I thought she could very easily do without me if she put her mind to it, and go off and make a life for herself. So I was relieved when she went back to college in Mayville. Some weekends she came back to town but I didn't see her as much as before. I told myself it was over, but when she asked, her voice breaking, "Oh, Jim, what about me? If you run out on me, what will I do?" and when I saw the lost look in her eyes, I could never bring myself to take that final step. . .

Nobody could remember such hard times. Almost every day, you'd hear about somebody's farm falling into the hands of a bank or being seized by the government for nonpayment of taxes, just as they'd done to our farm in Hunter in 1925. Farmers were getting less than twenty-five cents for a bushel of wheat, seven cents for a bushel of corn, a dime for a bushel of oats, two-and-a-half cents a pound for hogs and beef. Who could pay creditors at those prices?

The migrant army we'd first seen in Maddock had swollen to a flood of homeless, jobless men. Fessenden's hobo jungle had spread along the railroad tracks like fungus. There were packing crates, burlap-and-tarpaper shacks, lean-tos, wrecked cars, cardboard shelters; and those without even these slept on the hay at the stockyards. The clothing grew familiar: the filthy old sheepskin jacket, the buttonless suit coat, the pair of cast-off trousers out at the knee or in the seat, the patched tennis shoes, the mismatched canvas work gloves. Too familiar to a county coroner.

Abner Fryer, unemployed, no fixed address, being first duly sworn in at the coroner's inquest, testified as follows:

Q. "You were with Willie Carlson at the switchyard. Will you kindly state in your own words just what happened to Willie?"

A. "We'd been sleeping down in the stockyards and we went up to town and when I got to the depot, Willie left me. I stood there for a little while. . . . After a while Hadley came and told me that Willie was hit and I went back and found him under the freight cars and pulled him out. Another fellow came by and asked me if Willie was dead and I said, 'I don't know.' This other fellow was a yard bull, you know, a railroad policeman, so I walked away. . ."

Q. "Did he appear to be dead then?"

A. "Yes. I called him by name and he didn't answer. . . . He was all bloody. . . . I mean you could see Willie'd got crushed between two freight cars. Maybe he was trying to get across the tracks. . . . I didn't see any bottles or sign of drinking. . ."

Jim:

Willie Carlson's parents, farming people from Wisconsin, came to claim his body. They'd needed money to pay the mortgage and sent Willie off to ride the rails and look for work. As farm prices caved in, there were more and more Willies. . . . The automobile, too, played its role in keeping doctors busy. It had promised to be such a wonderful tool of freedom, whirling people where they wanted to go at exciting speeds. Now we were seeing more of its destructive side. As coroner, I discovered how many drivers and passengers were getting killed and maimed in county accidents. Somebody would tip over in his Ford ("fracture of skull and four left ribs, crushing internal injuries") or his car be struck by the Great Northern's Passenger Flyer ("must have been blinded by the sun's glare at crossing"). Or some young couple, just starting out in life, would park on a cold night to smooch ("it is thought fumes escaped from the heater of the car, filling the interior with deadly carbon monoxide gas"). There were the old, familiar causes of sudden death too, the farm accidents, strokes, heart attacks.

If the deaths of Gertrude Krause and Willie Carlson were tragic, what happened to Sven Gustafson was worse. It scared me so bad I began to despair of being able to hang on in Fessenden and thought of taking Anne and the five kids into Fargo, now a city of thirty thousand. Dr. Baillie of Hunter had gone there and I knew he'd help me to start a new practice. I was going to have to do something, and soon.

I'd known Sven—a tall, gray, upright Norwegian farmer of sixty-nine—for a good many years. He'd married a Maddock girl and we knew her family too. For years they'd been childless and then late in life they'd been blessed with a son—young Nels had just turned seventeen. Sven was hardworking, respected. People were to ask and ask, "What made him do it?" We'd never know, but suddenly losing one's farm no longer seemed the worst that could happen.

As I pieced it together, Sven had been brooding over the possible loss of his heavily mortgaged farm. He'd spent his whole life building it up. The day before, he paid off a neighbor for some help and told him, "That's the last three dollars I have in this world." At nine-thirty the next morning, an oil dealer from Hamberg named Liudahl knocked on the back door of the Gustafson house. He wanted to see if Sven wanted any tractor fuel. His creditors later claimed that Sven's farm was not in danger of foreclosure. Maybe this was true. But a man like Sven might think otherwise. He was terribly conscientious. Several witnesses said he himself feared he'd lose it. What we did know was that up until six o'clock that morning, life on the Gustafson farm seemed normal. All three of them—Sven, the wife and Nels—were seen by neighbors going about their chores just as on any other day. Even several hours later, I found everything at the barn and chicken house in perfect order.

So did Liudahl. He kept knocking. Nobody came. He tried the screen door. To his surprise, there was Gustafson seated in a chair across the kitchen table. Liudahl greeted him, then saw the gun in Sven's hand and the dried blood running down the side of his neck and heard the buzzing of the bluebottle flies. He realized he was staring into the eyes of a dead man. Liudahl ran for help, returning some minutes later with George Wynne and his wife from the next farm. They found Sven's brother, Peter, looking around the barn and wondering where everybody was. He had come to help Sven shock some wheat and was just about to go into the house.

All four went inside together. They confirmed Sven was dead. They found the boy in his bedroom. Evidently, after chores he'd gone back to bed. He'd been shot in the back of the head while he slept. His eyes were closed. He must have died instantly. There was no sign of Mrs. Gustafson. They found her in the doorway of a small storage room off the kitchen. I surmised she had been separating the cream when she heard one or more shots. She lay face down in the doorway. From her wounds, Sven appeared to have first shot her in the hand, which perhaps she raised to ward off the blow or to try and stop him. This bullet pierced her shoulder, too. She must have fallen then. Sven fired the fatal bullet into her heart after she'd already hit the floor.

For days after that, my hands would start shaking for no reason. I'd find myself emptying whiskey glasses at a single gulp, absently wiping my chin with my hand. I suppose, like so many were doing now, I despaired and drank, and drank and despaired. I'd think: if patients would come. . . pay their bills. Where were they? Sometimes it seemed as if all I was getting were the child cut up in the mower, the body left hanging until the birds got it, the brutally beaten drunkard's wife, the face kicked in by a horse, the scalded child, the injured thresher, the legless corpse on the railroad tracks, the family whose auto was struck by the train engine, the half-starved old settler dying alone on his farm. I knew some of the patients who had stuck with me were saying, "Even if you can smell whiskey on his breath, he never makes a wrong diagnosis." I clung to the hope there was some truth in it, that my understanding of the human body and its ills had not totally failed me.

Growing Up During the Depression

When I think of North Dakota now, I think of a hot, dry morning in July. Shortly before noon. Hazy with heat, a perfectly blue sky. The thermometer that day hit a hundred and twelve in the shade. We were shocking wheat on a forty-acre piece of land between the railroad embankment and the road that ran in front of the Nielsen farm. Jens and Chris had done the cutting with binders and left, and the girls, after shocking all morning, had gone back to the house too. I was right at the end of a row of shocks and was set on staying to finish it before lunch.

I was hurrying and I got dizzy. All of a sudden I fell backwards to the earth. The sky seemed to jump on top of me. It went round and round. I passed out. When I came to, I opened my eyes but I didn't move. I'd collapsed sort of half in the wheat stubble and half in the prairie grass by the railroad ties. It was that bluish-green buffalo grass that grows in little tufts. Not high grass. Crickets were chirping, a shrill, loud ringing in my ears. There were a lot of grasshoppers jumping about and one or two were sitting on my face. And I was lying there and I thought: how perfectly comfortable this is, nice and clean with the smell of the earth and the grass and, I guess, a little creosote from the railroad ties for the Great Northern train that went by there every day. After a while I recovered enough from the heatstroke or whatever it was to go find the water jug. I sipped that cold water until I was all right. But my whole feeling lying there in the stubble and the buffalo grass with grasshoppers crawling all over was, you know, this is my ambience—I was really at home there. I felt kind of childish euphoria, everything was good, I felt so peaceful and at one with the earth.

These further conversations with Jimmy, oldest son of Anne and Jim, took place from twenty-eight to fifty-five years later and were recorded stenographically or on tape.

We all followed Dad to Fargo the summer of 1932. It was different from anywhere we'd ever lived—a city—and from the beginning it seemed all challenge and uphill. I didn't like it; my own mood was one of rebellion mixed with hundreds of wonderful memories of Maddock and Fessenden. My first recollection was entering the city's big high school. The natural thing was to go out for football. I was like anybody fighting a system, terribly shy, embarrassed, determined. I was always playing football or basketball. I played every night. The spring of that first year I got yellow jaundice. I stayed home with a terrible fever, coughing yellow bile. What a wonderful nurse Mother was! Always bringing me grapefruit, oatmeal muffins, fresh honey. No one came to se me. No one called. April to mid-May. The day I went back to school, nobody noticed I'd been away. I got mad. That summer I wanted to get out of Fargo. I'd pined all year for Maddock and the farming days with Jens Nielsen. As soon as school was out, I went back.

I was fifteen that summer and Jens paid me a full man's wage—five dollars a day—for the first time. During threshing he gave me my own bundle rack and horses and harness. This old Dane, how shall I describe him? Black hair, mustache, red cheeks, chewed Copenhagen snoose. I could swear like a trooper in Danish at the age of ten. But Jens was never in such a hurry that he would not stop, pause to show me and explain to me why things were done the way they were. The easy way and the hard way. I went there every summer from the age of ten all through our seven years in Fessenden. At first I was essentially an auxiliary force for the Nielsen girls, getting the cows in, digging for potatoes, chasing chickens, filling up water jugs, hauling lunch out with a gray mare named Nell and a buggy.

Those girls were really good to me. Rose was indestructible, Aggie quiet and strong like her mother, Edith beautiful and with a good deal of humor. And Chris could do everything. He had a mechanical aptitude his father lacked, whereas I don't think he inherited Jens's love of horses. But everything Chris did, he did at a trot; he was always in a hurry. My folks got rid of me for practically the whole summer. I went over there and became the Nielsen's kid.

By then, Jens was running one of these great big North Dakota farms, though he always stayed with horses. There were three large horse barns, a lot of buildings and, as was common in those days, a well with a windmill, and watering troughs scattered around. The houses—all houses on farms around Fessenden and Maddock—were very simple, frame, many just one-story. Very few trees. The planting of windbreaks as part of the soil conservation program came after FDR. All the barns were red and all the roads were dirt. Nobody used fertilizer; they just kept using up the soil by planting wheat year after year.

At his peak, Jens had ninety-two horses at one time. Maybe a third were not pure-bred Clydesdales, but they had a lot of Clydesdale in them. He also bought range horses at auctions that were at least three-quarters just old mustang, probably bred on big ranches out West. When you harnessed a horse in the morning, you'd curry him first, clean the harness and then put this soft pad on his neck. Jens went to great effort to show me just how to drape a harness over my shoulder, how to approach the horse, and how to use my weight to get it up on him. I also helped Jens break in horses, usually in the fall plowing. We used different hitches, but most often we plowed with five horses, two in the lead and three so-called wheelhorses behind. And what we'd do is take one of these young mustangs and put them between two horses. They learned quickly. They had no choice. There's nothing that settles a horse down like pulling a plow.

Before harvesting, in the spring and early summer, we'd go over all the wagons, pulling the wheels and greasing them, tightening up the iron rims, wheelwright work, repairing all the harness, putting new handles on the broken pitchforks. Jens had a blacksmith shop with forges, a harness shop with leather and a general tool shop; it was exciting.

On rainy days the bunkhouse was a marvelous place, because the threshing crews would play cards all day long, and sing and play harmonicas and guitars and tell stories about St. Louie, Minneapolis, Winnipeg, Tulsa. They drank hot water, a little butter, cinnamon, sugar and raw alcohol. No matter what the ingredients, they always called it schnapps. There were two brothers, Shorty and Slim. Everybody knew Shorty because he was a big runner of alky down from Canada. He'd be missing for a day or two now and then. He brought his hootch back in silver gallon cans and he'd take them out and bury them in the tree claims. Every so often Shorty would get arrested and serve a little time for bootlegging. But he was a real hard worker for Jens. One carload, all from Oklahoma and Texas, always came in an Essex, and Tex, who drove it, and the other guys, always had musical instruments. I'd say most of the crew were second- or third-generation immigrants. An awful lot of them were Scandinavian or German farm boys who still spoke with an accent.

Once threshing started, we got up at four o'clock. The bunkhouse was simple. Straw ticks and a sheltered place out back with buckets and barrels of cold water and tin basins on a long table to wash your face in. I don't remember anybody having a toothbrush, but I suppose they must have. Saturday night, anybody going into town would be out there stripped down and scrubbing away, taking a bath in a bucket of water. Some were cleaner and some weren't. Mr. Thomas, the head of the threshing crew and a great big snoose-chewing man from Minneapolis, was famous for never having been known to take a bath. A lot of the threshers shaved every night, I suppose because dust and chaff could catch in a beard.

We'd get called and all stumble out of the bunkhouse and splash cold water on our faces. All farm people bought their clothes at the same place, so everybody looked alike: bib overalls, a long-sleeved Sears, Roebuck work shirt and a bandanna around your neck to keep the straw out. And an old battered hat of some kind. Why we didn't all die of lung disease I'll never know, as we worked in great clouds of dust. After washing, we'd all go down to the barn, take our horses out and water them, then bring them back and feed them hay and oats and put the harness on them. The harness had to be kept clean and properly adjusted. Jens fed his horses just twice a day, as we did later when I joined the U.S. Cavalry.

Breakfast was around five, five-thirty. Oh, gosh, it was huge. Great piles of eggs, bacon, sausage, thick slices of buttered home-made bread, oatmeal, potatoes, great mugs of coffee. There were two cook cars, one to cook in and one to eat in, great big things that got pulled around by horses in those days. They were usually in a farmyard, never at the threshing machine.

When threshing began each morning, the first thing you heard was the "Tooo-o-t!" of the steam engine. There was also a special toot for the Tankee, kind of a signal that told him, "Hey, our water's getting low! Get a move on!" There was a certain toot for lunch, and so on. A lot of tooting went on. In the evening, all of a sudden you'd hear this long whistle and then you'd hear the separator clank to a halt. Quitting time was determined by the straw's moisture content, the dew. During harvest, you often saw the sun setting late, maybe nine o'clock at night, and a threshing machine just shutting down. A good crew kept going as long as it could. The work ethic was unbelievable. There was a real tempo about a threshing machine.

The man who ran it all was the separator man, Mr. Thomas. He stood up on top, a great, towering man. He was the one who felt the grain, looked at the sun, smelled the wind, and said, "All right, boys, start." He made the decisions. He was always going around with a long-snouted oilcan, oiling, listening, adjusting, tuning. He had little walkways that ran all over the machine, and all day long he was watching every little piece of it. He was the boss.

Another man ran the steam engine. There was Tankee. And you had a Straw Monkey. Here you'd have this threshing machine spouting out straw and building up a huge pile. It was a real art. They'd consider the wind. They'd take a break and reposition the machine. Or Mr. Thomas would turn it up higher. Or he'd move it here and there. Pretty soon you had these great big straw stacks. The Straw Monkey would bring his rack and two horses and he'd back it in and Mr. Thomas would direct the blower so it would blow straw onto his rack. Then he'd take it and come back to the other end and shovel it into the steam engine.

It took a big crew. Besides them, you had twelve bundle haulers, two spike pitchers in the field and two at the machine and the grain haulers. Maybe thirty or forty men. Jens was always a grain hauler; they took the wheat into town to the elevator. Jens was the owner, the real boss. He took great pride in his horses, and during threshing he had four chestnut horses hitched to his wagon. Two were huge big mares—Bess and Jewel. He also drove them on his bobsled in winter. Anytime Jens drove anyplace, he drove Bess and Jewel.

What you did, you drove your team out to fields where a binder had cut and bound the harvested wheat into bundles for threshing. The bundles were piled to dry in neat shocks which were all lined up in rows, sometimes reaching miles and miles, all the way to the horizon. You'd go to one end of a field and hop off and tie the reins up and get your horses pointed down a line of shocks. You always got on the upwind side. You had a three-tine pitchfork and at first you threw the bundles into the rack helter-skelter. When you had four or five feet of them, you'd start building a load by sticking the pitchfork in the bottom of a bundle so the heads of grain pointed away from you. You'd then reach up and start building a very straight wall with these bundles. The pride of every good bundle pitcher was the ability to build his load up to incredible heights. Once you got a load on, you'd climb up, stepping on the rump of one of your horses. You left a kind of ladder. You'd climb up to where you'd tied your reins on the front pole. And you'd sit way up there and drive back to the machine and take your place in line.

Threshing rigs differed, but the principle stayed the same. After you'd cut the standing crop and gathered it up, you needed to thresh the seed from the stem, separate out the chaff, return the straw to the ground and collect the grain in a wagon to haul it to the elevator in town to sell. As you drove up to the rig, you pulled in beside the separator, lining up your rack exactly parallel with an exposed moving platform—it was on link chains and had forty-inch hardwood boards which connected these chains. When the front end of your bundle rack was far enough, you started dropping bundles on this moving platform. At once, the bundles were carried toward a revolving cylinder armed with beaters; in the front were about five or six axles with little sharp knives. Going like crazy. If anybody fell into it, they'd quickly be chewed up by those knives. It happened occasionally. After it was chopped up, the wheat passed over a series of shaking screens, sieves and rollers that further separated the chaff, straw and grain as blowers carried off loose strands. At last the straw and chaff were blown into the air in clouds of dust, setting into a big stack. About halfway down the rig, there was a chute where the stream of grain came out. Here Jens and the other grain haulers backed their wagons in. You always had one filling up and another ready to back in.

Mr. Thomas had to decide whether to put his spike pitchers in the field to make loading go faster, or keep them at the machine. It was a forty-four-inch Case separator, one of the bigger rigs of the time. It took twelve bundle racks to keep it supplied with grain; six kept going in a circle on each side. You'd usually arrive with a full load just as the wagon ahead of you was moving in to start unloading, so there was time to talk and relax and joke, and there would be a flash of white teeth in the dust-blackened faces. Sometimes Mr. Thomas shut down the rig to adjust something. Workhorses on a farm, you didn't have to stay with them, you know. They were so tired most of the time. Oh, they were well fed—big, fat, glossy animals You could jump down from the rack onto their big broad hindquarters and from there to the ground. They were strong.

I loved it. Every minute of it. I loved every minute of the farm work. I suppose I regarded it as high adventure. Absolutely.

We all expected to stay in farming all our lives. None of us knew there was any other world. How would you know? I can remember as a kid of thirteen or fourteen, working in a wheat field, and seeing a Model T go by. They'd graded the road by Jens's farm and put gravel on it; that made it kind of a highway then. At that time it was well known that there were just thirteen miles of pavement in the whole state of North Dakota, and they ran from Fargo to West Fargo. After that, you hit dirt. Right beyond that road was the railroad track. And it was all open. Not a tree. I stopped and watched that car go by and I can remember thinking, "I wonder who those people are? Where are they from? Where are they going?" But I never thought of going off myself. It just never came to mind.

You expected it all to go on forever. Just picture those sunbaked plains, dirt roads with clouds of dust as the occasional horse-drawn cart or Model T passed by. You'd find a small family farm every half or quarter section, with cows, pigs and chickens. It was a rural way of life still mostly set by the three-mile-per-hour gait of the horse. At the Nielsens in 1932 we still used kerosene lamps. It was already going, though. By 1937, the last summer I threshed, joining a crew out at Steve Litke's farm in Fessenden, they were already using tractors and everybody was getting mechanized. It wasn't the same any more.

Jens kept his horses to the last. But the day of the steam engine was just about over. Keeping those great monster threshing machines going was getting too difficult. Tractors were appearing, and everybody was talking about the new combines that could cut the wheat, gather it up and thresh it, separate out the chaff and straw and collect the grain in a hopper for delivery to a truck. With night lights, working round the clock, one of the combines could thresh up to a hundred and twenty acres a day. I suspect Mr. Thomas died. The next farm to the south bought a twenty-eight-inch separator in Minneapolis, which was powered by a tractor. It was a lot more efficient and needed fewer men.

That summer of 1932, things began to hit bottom for Jens. Spring wheat, which had been two dollars and seventy-six cents a bushel in 1920, that fall dropped to twenty-six cents a bushel, the lowest wheat price ever. When I went back the summer of 1933, it was down to twenty-five. Mama Nielsen was trying to sell as many eggs as she could, but they had dropped from thirty cents to thirteen cents a dozen. Neighbors were losing their farms. North Dakota was drying up.

Back in Fargo, a big change came in my junior year. I got invited to a party and met Connie Taylor, who became my girl, and Ross Porter, soon to be my best friend. They were in a group of the most wholesome kids you can imagine. Having cracked the ice, I entered a real social whirl. There were a lot of dancing parties and you got printed programs with twelve dances. For days it was a big game for the girls. Connie would meet Lavonne and Joyce and plan who would get what dance, the waltz, the foxtrot, the Charleston, the Big Apple. The boys only found out that night. I earned a letter on a state-championship football team, got elected president of the Sportsmanship Club, played intramural basketball and trumpet in the band. Yet somehow it was an agonizing year, maybe because I was at the height of adolescence.

Many times, I caddied for Father out at the Country Club. He golfed with a group of friends and of course he was very good at it. He'd ask me in a kind of pleasant way. We had kind of a pleasant relationship. The question of saying no to him never entered my mind. If Father said he wanted you to go caddying, I just assumed that was what you did. But you know, I can't remember any football game, any event, any time, in my life, that I participated in that he came to. I don't think the thought ever occurred to him. He saw hunting essentially as a social event with his friends. I think he and Mother had a lot of fun. I can remember them going off golfing and duck hunting and partying and camping. And Mother was always interested and concerned, but there were limits to what she could do.

So it was Jens Nielsen who played the major role in my growing up, along with maybe Mother's brothers-in-law in Iowa, Fay and Victor Collins. We'd usually drive down to Iowa just after school was out, and Victor and Helen, or Fay and Mary, would drive me back to North Dakota by the middle of July. I'd be in Iowa for haying and then get up to Maddock in plenty of time for the harvest. All three of these men gave you the feeling that if they were going to explain something to you, or do something with you, or show you something, you really had their undivided and very genuine attention.

I can't remember Father once showing me anything, telling me anything, explaining anything to me. I had the feeling he used me, that was all. That second year in high school, his health was beginning to deteriorate. This was starting to cast its shadow. Mother was essentially sad and troubled. I sensed sadness and trouble in my own life too, but I probably couldn't have identified the cause.

The summer of 1934, Jens lost the whole wheat crop. There was nothing to thresh. What happened was a great cloud of grasshoppers dropped out of the sky one day, millions of them, flying low, some almost touching the ground, others flying so high they blotted out the sun. They were great big insects, three or more inches long, with black, leathery wings and a smaller set of bright-red or purple wings behind. Wide greenish-yellow backs, but like no grasshoppers anybody'd ever seen before. More like locusts. We'd been hearing about grasshopper attacks since 1931, when the first gigantic swarms appeared. They'd descend on a farm and not only eat all the leaves and stems they could find, they'd chew the binder twine off a sheaf of grain or gnaw on the wooden handles of farm tools. It was terrible. For three years the Nielsen farm escaped. Then they came.

Millions settled, just everywhere, with a crackling roar like burning fields of stubble. Worst was the high-pitched piercing noise they made, a kind of nightmarish grasshoppers' song the scientists later told us was stridulation, made when millions of hind legs rubbed against the whirring wings. One minute it was quiet and the next you couldn't hear a thing for that damn noise and they lit on the wheat, in the grass, on the roof of the house and the barns, on the walls, the posts, the trees. A few got into the house before Mama Nielsen got all he windows shut. There was no escaping them. All of a sudden, right in the middle of the day, the whole farm was covered with a greenish-yellow, crawling, moving mass, and all the while that terrible shrill, screaming sound. Aggie and Edith were caught outside and came shrieking into the house, brushing grasshoppers out of the hair and off their clothes. We ran to round up the stock, and so many insects kept coming, it was soon like a dark blizzard of grasshoppers. They ate every green, growing thing on the farm right down to the ground in a matter of minutes—all the wheat, the oats, the barley, the pasture, they even stripped the leaves off all the trees. In no time they cleaned out several hundred acres for Jens. We and the neighbors tried to fight them with bonfires but it was no use. We gathered a little grain the grasshoppers spared and tried to thresh it but got practically nothing. It was worthless.

Once the grasshoppers left, the bare, stripped earth started to blow. What the grasshoppers hadn't eaten, the dust blew away. Afterward I walked around the farm with Jens. Every place we looked, there was nothing but Russian thistles blowing across bare dirt. Jens was just about a broken man.

"I plowed it all and I seeded it and I waited for it to grow and what did I get?" His voice cracked and his eyes were bright with tears. "Nothing." I walked behind him as the Russian thistles blew across the dry, cracked ground, and Jens went along, pulling them away from his fences; wherever they caught on the barbed wire, they stopped the blowing dust and topsoil and it would drift up and catch more thistles. Just wind and dust and Russian thistles. The Nielsen farm was starting to disintegrate. Oh, there was enough to eat, but the great horse barns stood almost empty. In time, of course, soil conservation, crop rotation, tree belts, rural electrification, all the New Deal programs under FDR would rescue towns like Maddock and Fessenden from the Dust Bowl. By then, farming would be mechanized and the days of horses gone forever.

From Chester Zumpf, interviewed forty-eight years later; a farmer in 1934, now a retired implement dealer in Fessenden:

"Those grasshoppers didn't all hatch here. They hatched in Colorado and South Dakota and all the way down the line. The wind blew 'em here. They came in with, maybe, ten thousand feet of altitude. You'd look up and you could hardly see the sun. It was just gray. Solid with grasshoppers. And then if the wind let up or towards evening, they'd come down to feed. When they came down it was just like a terrible storm of grasshoppers. When they got through eating, there wasn't anything around. You could walk on 'em three deep. They'd jump at you but didn't bother you any. They were the same as locusts, big fellows, with black in their wings, and they could fly almost like little sparrows. If there was linseed paint, they'd eat it right off a house or a barn. Fields that had a little growth on them, maybe six, eight inches, well, when those grasshoppers finished, in a couple hours' time, it was eaten right down to the ground. Nothing left."

Further comment from Agnetha Nielsen Bergsgaard, interviewed on her farm near Maddock forty-eight years later:

"We found out what the Depression was when we were living in

Wisconsin. I hadn't been married so long and my husband was working for the Case company. One day, he came home and said, 'I think we'd better start back home for North Dakota. They're laying off people to beat the band down there.' So we came back and rented this farm. We'd saved two thousand dollars in Wisconsin and we put half of it in the Flora bank and half of it in the Hamberg bank. We lost it all. Both banks failed and we lost all that money. So I had chickens. I got seven cents for a dozen eggs. I sold cream, eleven cents a pound for butterfat. We milked and separated that, you know, and sold it at the creamery. And we raised our own garden. So we could keep on living. That's what my husband said, 'We get back out on the farm we can always get some food anyway.' And that's what we did. We could always eat.

"We bought the farm in 1938. We were sitting at the table, making out the contract and we bought it for thirty-four dollars an acre. Afterwards I made some coffee and we turned on the radio and heard President Roosevelt say we might have to go to war with Germany. 'Don't be surprised if we're in it,' he said. We had grain bins all full of wheat. We hadn't sold it, because it was only about seventy-five cents a bushel. For a couple of weeks, it went up to seven dollars a bushel. Well, we made a payment on the farm for three hundred and forty dollars that fall, and the next year we paid the rest of it and we've owned it ever since. Without a cent of anything against it. This spring, two quarters of land down the road here sold for eight hundred dollars an acre. And it was just hills and sloughs I don't know why in the world anybody would pay eight hundred dollars for that poor a land. . ."

Jimmy:

In the spring of 1935 I was named "Representative Senior," the biggest honor Fargo High School had to give. It stunned me. I'd been genuinely shy and unassuming and had no feeling of being outstanding. I went to school most days wearing a rather peculiar light tan corduroy jacket Mother had made on her sewing machine. The award was given at a big general assembly in the school auditorium. Claire Putz, the girl chosen, and I were called to the stage.

Everybody cheered and applauded. The commander of the local American Legion post gave us each a check for ten dollars. Mother was there; she'd been invited especially and was pleased as punch. I don't remember Father taking any notice of it at all.

Hardship

By David Martinson

Never could they afford enough
To pay a nickel for a cup
Of coffee. Going in to town
With horses in the depression,
My mother then a girl (under
The weight of a buffalo robe,
Her legs bound up in woolen socks)
Winter glazed gawked in the wonder:

The window of Wangenstein's Store.
Perhaps some peppermints, but more
Likely than not, flour and salt,
Some wicks to hold onto the light
That glowed in the kerosene lamp,
Or oil for the sewing machine
On credit until such time
The check from the creamery came.

Enough to go to town and look
And wish and wait and window shop
And know her folks were better off
Than those sold out by the sheriff.
They had the common sense to save
On things, to know that waste was sin.
My mother grew in the hardship,
Never expecting it to end.

Velva, North Dakota: 1946 and 1964

After explaining in Not So Wild a Dream *why he had not returned to his native state, television commentator Eric Sevareid did return. His explanation and the account of his visit provide contrasting perspectives on a North Dakota town.*

VELVA, NORTH DAKOTA, 1946

By Eric Sevareid

The small brown river curved around the edge of our town. The farmers plowed close to its muddy banks and left their water jugs in the shade of the willows. There is not much shade in the northern sections of North Dakota, nor is there much shelter in the wintertime. Even as very small children we could sense the river's life-giving nature and meaning to the farmers, to us all. By December, despite the river's current, the men could cut ice blocks three feet square, to be stacked and layered with sawdust in the shed behind Moose's general trading store on Main Street, against midsummer when the hot winds came across the prairie, a time when the milk seemed to sour just a few minutes after you had milked the cow, when you couldn't even be sure of the butter kept in the well.

Velva (nobody seemed to know where the name came from; I suspect it made a pretty sound to the wife of some early settler) was only one of various villages strung upon the river's wandering length, but naturally we felt we exercised particular rights of possession over its flowing. On the red-painted wooden bridge, leading into the "city park" was mounted a large sign bearing a white star and the words in block letters: "Star City on the Mouse." This led also to the baseball diamond and the swimming hole just beyond. Sometimes the team from a village like Voltaire would come to play our men, and I can still remember my own feeling of proud generosity when, after the game, the Voltaire team would hurry to our swimming hole, strip off their overalls, and slide down our mud slide in to the water, shouting, splashing, and shoving one another. Voltaire was only a few miles away, but it was bare and riverless. These men swam awkwardly with a great deal of thrashing and spitting. Only their forearms were burned a dark brown, and the face and neck down to the junction of the collarbone. The rest of their bodies was dead white in contrast. Grown men in those climates did not expose their bodies to the sun, and it was years before I saw adults with carefully nurtured "tans," acquired in leisure, not working, time.

Wheat. So far as Velva was concerned, wheat was the sole source and meaning of our lives, which were given in continuing hostage to the vagaries of this pewter-colored ocean that lapped to the thistle-covered roadbed of the Soo Line and receded in perpetually undulating billows as far as a child could see from the highest point, even from the top of the water tank. We were never its masters, but too frequently its victims. It was our setting and scenery. It was rarely long outside a conversation. On the mercy of the wheat depended the presence of new geography books in the red brick schoolhouse, a new Ranger bicycle from Montgomery Ward's, good humor in my father's face. Its favor or disfavor determined the size and mood of the crowd of farmers on Main Street Saturday nights, and was the reason Pastor Reishus in the Lutheran church prayed as frequently for rain as he did for our immortal souls. In good harvests it meant that hordes of itinerant workers, I.W.W.'s (which meant "I won't work," according to the businessmen of the town), hung around the poolroom and Eats Cafe, hunched like tattered crows on the hitching rails, spat tobacco juice at the grasshoppers in the dusty street, and frightened the nice women of the town so that they rarely ventured on Main Street in the evening time. Good harvest meant that my father would have to leave his office in the back of the little bank, remove his hard white collar, change to overalls, and, taking my older brother with him, go to help out on one of the bank's farms by driving the four-horse binder, while Paul, who was big for his age, would struggle with the shocks. Hired man or town banker, wheat was the common denominator of this democracy. It made all men equal, in prosperity or wretchedness. It meant that my father, the banker, was more of a confessor than the Catholic priest. His office was connected by a door to the town library for a time, and I could slip in among the bookshelves on days when the library was closed to everyone else. Sometimes I looked through the keyhole. I remember times when I would see a gaunt, unshaved Norwegian farmer sitting before my father's desk, staring down at his blackened nails, speaking to my father with painful difficulty about the locusts or the reaper which broke its axle on a rock, and sometimes, with more difficulty, about his wife who had gone sick again. Those were in the bad times. Those were the days when my father could not eat much at supper. Those were the days when a buggy would drive up to our house after supper and my father and a wheat farmer would sit on the porch talking in low tones, with long periods of silence, until after we children fell asleep upstairs. Wheat was our solace and our challenge. My mother, who came from a green and pleasant city in the distant, mystical

East—in Iowa—feared and hated it. My father simply met the challenge without emotion, as a man should, and grappled with it as well as a man knew how. In the end he lost. It ruined him.

North Dakota. Why have I not returned for so many years? Why have so few from those prairies ever returned? Where is its written chapter in the long and varied American story? In distant cities when someone would ask: "Where are you from?" and I would answer: "North Dakota," they would merely nod politely and change the subject, having no point of common reference. They knew no one else from there. It was a large, rectangular blank spot in the nation's mind. I was that kind of child who relates reality to books, and in the books I found so little about my native region. In the geography, among the pictures of Chicago's skyline, Florida's palms, and the redwoods of California, there was one small snapshot of North Dakota. It showed a waving wheatfield. I could see that simply by turning my head to the sixth-grade window. Was that all there was, all we had? Perhaps the feeling had been communicated from my mother, but very early I acquired a sense of having no identity in the world, of inhabiting, by some cruel mistake, an outland, a lost and forgotten place upon the far horizon of my country. Sometimes when galloping a barebacked horse across the pastures in pursuit of some neighbor's straying cattle, I had for a moment a sharp sense of the prairie's beauty, but it always died quickly away, and the unattainable places of the books were again more beautiful, more real.

VELVA, NORTH DAKOTA, 1964

By Eric Sevareid

My home town has changed in these thirty years of the American story. It is changing now, will go on changing as America changes. Its biography, I suspect, would read much the same as that of all other home towns. Depression and war and prosperity have all left their marks; modern science, modern tastes, manners, philosophies, fears and ambitions have touched my town as indelibly as they have touched New York or Panama City.

Sights have changed: there is a new precision about street and home, a clearing away of chicken yards, cow barns, pigeon-crested cupolas, weed lots and coulees, the dim and secret adult-free rendezvous of boys. An intricate metal "jungle gym" is a common backyard sight, the sack swing uncommon. There are wide expanses of clear windows, designed to let in the parlor light, fewer ornamental windows of colored glass designed to keep it out. Attic and screen porch are slowly vanishing and lovely shades of pastel are painted upon new houses, tints that once would have embarrassed farmer and merchant alike.

Sounds have changed; I heard not once the clopping of a horse's hoof, nor the mourn of a coyote. I heard instead the shriek of brakes, the heavy throbbing of the once-a-day Braniff airliner into Minot, the shattering sirens born of war, the honk of a diesel locomotive which surely cannot call to faraway places the heart of a wakeful boy like the old steam whistle in the night. You can walk down the streets of my town now and hear from open windows the intimate voices of the Washington commentators in casual converse on the great affairs of state; but you cannot hear on Sunday morning the singing in Norwegian of the Lutheran hymns; the old country seems now part of a world left long behind and the old-country accents grow fainter in the speech of my Velva neighbors.

The people have not changed, but the *kinds of* people have changed: there is no longer an official, certified town drunk, no longer a "Crazy John," spitting his worst epithet, "rotten chicken legs," as you hurriedly passed him by. People so sick are now sent to places of proper care. No longer is there an official town joker, like the druggist MacKnight, who would spot a customer in the front of the store, have him called to the phone, then slip to the phone behind the prescription case, and imitate the man's wife to perfection with orders to bring home more bread and sausage and Cream of Wheat. No longer anyone like the early attorney J. L. Lee, who sent fabulous dispatches to that fabulous tabloid, the Chicago Blade, such as his story of the wild man captured on the prairie and chained to the wall in the drugstore basement. (This, surely, was Velva's first notoriety; inquiries came from anthropologists all over the world.)

No, the "characters" are vanishing in Velva, just as they are vanishing in our cities, in business, in politics. The "well-rounded, socially integrated" personality that the progressive schoolteachers are so obsessed with is increasing rapidly, and I am not at all sure that this is good. Maybe we need more personalities with knobs and handles and rugged lumps of individuality. They may not make life more smooth; more interesting they surely make it.

They eat differently in Velva now; there are frozen fruits and sea food and exotic delicacies we only read about in novels in those meat-and-potato days. They dress differently. The hard white collars of the businessmen are gone with the shiny alpaca coats. There are comfortable tweeds now, and casual blazers with a touch in their colors of California, which seems so close in time and distance.

It is distance and time that have changed the most and worked the deepest changes in Velva's life. The telephone, the car, the smooth highway, radio and television are consolidating the entities of our country. The county seat of Towner now seems no closer than the state capital of Bismarck; the voices and concerns of Presidents, French premiers and Moroccan pashas are no farther away than the portable radio on Aunt Jessey's kitchen table.

The national news magazines are stacked each week in Harold Anderson's drugstore beside the new soda fountain, and the excellent Minot Daily News smells hot from the press each afternoon.

Consolidation. The nearby hamlets of Sawyer and Logan and Voltaire had their own separate banks and papers and schools in my days of dusty buggies and Model T's marooned in the snowdrifts. Now these hamlets are dying. A bright yellow bus takes the Voltaire kids to Velva each day for high school. Velva has grown—from 800 to 1,300—because the miners from the Truax coal mine can commute to their labors each morning and the nearby farmers can live in town if they choose. Minot has tripled in size to 30,000. Once the "Magic City" was a distant and splendid Baghdad, visited on special occasions long prepared for. Now it is a twenty-five minute commuter's jump away. So P. W. Miller and Jay Louis Monicken run their businesses in Minot but live on in their old family homes in Velva. So Ray Michelson's two girls on his farm to the west drive up each morning to their jobs as maids in Minot homes. Aunt Jessey said, "Why, Saturday night I counted sixty-five cars just between here and Sawyer, all going up to the show in Minot."

The hills are prison battlements no longer; the prairies no heart-sinking barrier, but a passageway free as the swelling ocean, inviting you to sail home and away at your whim and your leisure. (John and Helen made an easy little jaunt of 700 miles that weekend to see their eldest daughter in Wyoming.)

Consolidation. Art Kumm's bank serves a big region now; its assets are $2,000,000 to $3,000,000 instead of the $200,000 or $300,000 in my father's days. Eighteen farms near Velva are under three ownerships now. They calculate in sections; "acres" is an almost forgotten term. Aunt Jessey owns a couple of farms, and she knows they are much better run. "It's no longer all take out and no put in," she said. "Folks strip farm now; they know all about fertilizers. They care for it and they'll hand on the land in good shape." The farmers gripe about their cash income, and not without reason at the moment, but they will admit that life is good compared to those days of drought and foreclosure, manure banked against the house for warmth, the hand pump frozen at 30 below and the fitful kerosene lamp on the kitchen table. Electrification has done much of this, eased backbreaking chores that made their wives old as parchment at forty, brought life and music and the sound of human voices into their parlors at night.

And light upon the prairie. "From the hilltop," said Aunt Jessey, "the farms look like stars an night."

Many politicians deplore the passing of the old family-size farm, but I am not so sure. I saw around Velva a release from what was like slavery to the tyrannical soil, release from the ignorance that darkens the soul and from the loneliness that corrodes it. In this generation my Velva friends have rejoined the general American society that their pioneering fathers left behind when they first made the barren trek in the days of the wheat rush. As I sit here in Washington writing this, I can feel their nearness. I never felt it before save in my dreams.

But now I must ask myself: Are they nearer to one another? And the answer is no; yet I am certain that this is good. The shrinking of time and distance has made contrast and relief available to their daily lives. They do not know one another quite so well because they are not so much obliged to. I know that democracy rests upon social discipline, which in turn rests upon personal discipline; passions checked, hard words withheld, civic tasks accepted, work well done, accountings honestly rendered. The old-fashioned small town was this discipline in its starkest, most primitive form; without this discipline the small town would have blown itself apart.

For personal and social neuroses festered under this hard scab of conformity. There was no place to go, no place to let off steam; few dared to voice unorthodox ideas, read strange books, admire esoteric art or publicly write or speak of their dreams and their soul's longings. The world was not "too much with us;" the world was too little with us and we were too much with one another.

The door to the world stands open now, inviting them to leave anytime they wish. It is the simple fact of the open door that makes all the difference; with it opening the stale air rushed out. So, of course, the people themselves do not have to leave, because, as the stale air went out, the fresh air came in.

Human nature is everywhere the same. He who is not forced to help his neighbor for his own existence will not only give him help, but his true good will as well. Minot and its hospital are now close at hand, but the people of Velva put their purses together, built their own clinic and homes for the two young doctors they persuaded to come and live among them. Velva has no organized charity, but when a farmer falls ill, his neighbors get in his crop; if a townsman has a financial catastrophe his personal friends raise a fund to help him out. When Bill's wife, Ethel, lay dying so long in the Minot hospital and nurses were not available, Helen and others took their turns driving up there just to sit with her so she would know in her gathering dark that friends were at hand.

It is personal freedom that makes us better persons, and they are freer in Velva now. There is no real freedom without privacy, and a resident of my home town can be a private person much more than he could before. People are able to draw at least a little apart from one another. In drawing apart, they gave their best human instincts room for expansion.

Depot at Alice, Medora, Slant Village, Motel, and Fort Berthold Reservation

DEPOT AT ALICE

By Richard Lyons

The depot at Alice was busy,
at least on the west end,
which was occupied by bank
swallows under the eaves
and one worried robin
on the window sill
that objected to anyone
banging around looking
at her two babies
with open mouths, demanding.
The swallows were busy
with their private schedules
of domestic traffic,
keeping their own passengers
satisfied in their clay bowls
like lopsided pottery.
Beyond this coming and going
ALICE stood, surrounded
by grass, looking
vainly down the blank tracks
for some reason to be.

MEDORA

By Richard Lyons

The Badlands are old lures of subdued color.
They carry a silent history of cattlemen and cows
And of native rites before them, slaughtered.
The hills slope rounded downward,
Like large and beautiful Mandan lodges
Colored with the faded blood of gone buffaloes.
Backed against one of these hills, the small town of Medora,
Named by a white pirate with primitive illusions,
Nestled for years, ignored, forgotten,
On the edge of the Theodore Roosevelt Memorial Park.
Its tallest structure was the lone chimney
Of DeMores' sacked and shattered abbatoir.
In 1952 there were four bars, a community hall,
And the Rough Rider Hotel and telephone exchange,
From whose balcony Tom Mix had leaped to Hollywood.
A coal stove at the end of a cold corridor
Heated the rooms if the doors were left open
But not if, like the hall toilet, the doors were closed.

All that has altered; the town is booming now.
On the way in along I-94 the oil rigs are drilling
And the oil pumps are bobbing in the Park.
Medora's filled with tourists and souvenir shops,
Imitating Wall Drug Store in South Dakota,
With bars, museums, restaurants, and the Badlands Motel
With never a vacancy in its paved corral.
The good-times sun shines like a gold seal
In the profitable and happy blue sky
Made to order from the good old formulas
Of romance and good business, the latter first.
In a small park bordered by a stone fence,
A quiet place, a shrine in all the hoopla,
A statue of the Marquis DeMores stands tall,
With a medallion of Medora, his wife, on the pedestal.
His right hand, holding a cape, points out from his hip;
He faces away, with her perspective,
From the large farmhouse across the railroad embankment
Like a dike and the Little Missouri River.
The house still holds up its heavy title of Chateau

The Marquis gave it to impress the sod-hut cowboys.
He knew Roosevelt but not the robber barons
of the Chicago slaughterhouses. He failed in his day.
He has been revived, like Theodore Roosevelt,
In cast metal, and he stands like a victor
With the large Chuckwagon Cafeteria behind him,
A double monument to his madcap spirit.

The old days are gone, the roughrider days,
The days of grime, sweat, thirst, and short lives,
The days of Old Four Eyes and the Marquis,
Of dusty greed blinded to beauty.
Here are the new days of air-conditioned rooms,
Of cocktails and souvenirs of plastic Indians,
Of photographing tourists all making the thing pay,
Of new Old Four Eyes and DeMores succeeding
Like the skillful carnies the old ones tried to be.
As I look upward here in this quiet shrine
Of reverent flagstones at the statue of DeMores,
His right hand waves a blessing on the town.

SLANT VILLAGE

By Richard Lyons

1
Having traveled the hard freeway
and maneuvered a narrow road
and the leaf-fringed footbridge,
I stand now weighted with credit cards
and calendars in this enclosure.
The light is as soft as pussywillow,
the earth ancient dust, powdered
by feet, most recently shod hard.
This is a fallout shelter
to sever me momentarily from the fissioning
bombardment of the twentieth century's
demand that I buy and smile
and be always concerned with odors
in which the earth is never mingled.
In the white light from the entrance
the half-log benches turn silver,
and the skylight slants a living square

of burnished sun in the dust,
antiquity's untouchable gold
at the end of the gone rainbow.

2
In this earth lodge now
the dusty household wealth
hangs behind a hard fencing,
imprisoned to protect it.
Free it, it would disappear
in a final white ravage.
It shows now in a brown silence—
the buffalo robe of the chief,
the dog travois and the horse travois,
the powdery earth flower and the lodge poles—
in the sandy light
of the sky transformed
through the square opening for smoke
above the half-burned logs of cold fire.
Within the tribal atmosphere
of a dark past in the dead air
only the nameplates are white.

3
Each lodge,
from the dark interior
of each lodge,
as one looks out,
holds, each one,
the Arc of the First Man,
the protected red symbol
wrapped in a circle of logs
(which makes in perspective
a rectangle
like the lined logs of the doorway,
the shape of the open entrance
and the closed world).
But the myth
or meaning
the circle holds
is withheld from us
now wrapped
in our own circles

of multiple stress,
whose lodges face inward,
each on its dark center
of competitive fire.

MOTEL

By Richard Lyons

1
The Chieftan Motel in Carrington
is all Indian in theme.
Its sign is supported with totem poles
no Sioux ever saw, nor Hidatsa,
but the totems hanging thereon
are universal: AAA, Best Western,
and Master Charge.

2
In the coffee shop
are beads and medallions
made in Hong Kong,
and a frieze of Indian portraits
looks down on a model tipi
and a bust of an Indian
mounted on the pie counter.

3
But in the hall
authentic artifacts
hold out beauty and mystery
behind glass,
from the beauty of Adolph Wisespirit's
buckskin shirt that Josephine Holy Spirit,
his wife, made
to the mystery of the White Riders,
an organization of local men
that Adolph belonged to
in the 20's and 30's,
since died out.
Adolph, to what pale purpose
in your beautiful shirt
did you ride in a crowd?

FT. BERTHOLD INDIAN RESERVATION

By Richard Lyons

The water rises on the Reservation,
covering the private past of the Indians,
even from the people of the Three Tribes.
On the high ground above the lake
tanks and towers rise
alien to the people of the Three Tribes.
But men on horses still ride
back and forth, back and forth
to keep the cattle bunched and moving
toward and into the rail corral,
heading off the obdurate calf
or the bawling cow, their horses
responding in preparation for a rodeo.

The water rises under Four Bears Bridge,
the northern gate to the Reservation
and to Four Bears Recreation Area,
with its Motor Lodge
planned by the Corps of Engineers
and now operated by Indians,
riding back and forth, back and forth
between the old and the new,
driving with an ancient will
into the American way of survival.
There is also a fine Indian museum
guarding the sacred arts and rituals
of the Indian way of living
beside a Standard Oil filling station.
Swimming, boating, water skiing
and picnics can be practiced by all
in the Indian area of white recreation,
and once a year, across the highway
from the lodge, the old joys
are recreated in a pow wow
high above the fat river,
with the old buttes looking on,
as Four Bears, Raven Chief, and Gray Prairie Eagle
dance again in the pure air.

The lodge is air-conditioned.
The young Reservation girls,
as clerks and waitresses getting used to clocks
on the border of two time zones,
wrestle with the American way of life
with stoic resignation,
suspended between two worlds.
It may be they have the best of both.

A Place Unvisited

By John Steinbeck

In late summer in 1960, John Steinbeck and Charlie, a French poodle, toured the United States in a pickup camper named Rociante (after Don Quixote's horse). Home again, Steinbeck put his and Charlie's experiences into his book, Travels with Charley. The chapter reproduced here records his experiences in North Dakota.

If there had been room in Rocinante I would have packed the W.P.A. Guides to the States, all forty-eight volumes of them. I have all of them, and some are very rare. If I remember correctly, North Dakota printed only eight hundred copies and South Dakota about five hundred. The complete set comprises the most comprehensive account of the United States ever got together, and nothing since has even approached it. It was compiled during the depression by the best writers in America, who were, if that is possible, more depressed than any other group while maintaining their inalienable instinct for eating. But these books were detested by Mr. Roosevelt's opposition. If W.P.A. workers leaned on their shovels, the writers leaned on their pens. The result was that in some states the plates were broken up after a few copies were printed, and that is a shame because they were reservoirs of organized, documented, and well-written information, geological, historical, and economic. If I had carried my guides along, for example, I would have looked up Detroit Lakes, Minnesota, where I stopped, and would have known why it is called Detroit Lakes, who named it, when, and why. I stopped near there late at night and so did Charley, and I don't know any more about it than he does.

The next day a long-cultivated ambition was to blossom and fruit.

Curious how a place unvisited can take such hold on the mind so that the very name sets up a ringing. To me such a place was Fargo, North Dakota. Perhaps its first impact is in the name Wells-Fargo, but my interest certainly goes beyond that. If you will take a map of the United States and fold it in the middle, eastern edge against western, and crease it sharply, right in the crease will be Fargo. On double-page maps sometimes Fargo gets lost in the binding. That may not be a very scientific method for finding the east-west middle of the country, but it will do. But beyond this, Fargo to me is brother to the fabulous places of the earth, kin to those magically remote spots mentioned by Herodutus and Marco Polo and Mandeville. From my earliest memory, if it was a cold day, Fargo was the coldest place on the continent. If heat was the subject, then at that time the papers listed Fargo as hotter than any place else, or wetter or drier, or deeper in snow. That's my impression, anyway. But I know that a dozen or half a hundred towns will rise up in injured wrath to denounce me with claims and figures for having much more dreadful weather than Fargo. I apologize to them in advance. As a sop to hurt feelings, I must admit that when I passed through Moorhead, Minnesota, and rattled across the Red River into Fargo on the other side, it was a golden autumn day, the town as traffic-troubled, as neon-plastered, as cluttered and milling with activity as any other up-and-coming town of forty-six thousand souls. The countryside was no different from Minnesota over the river. I drove through the town as usual, seeing little but the truck ahead of me and the Thunderbird in my rear-view mirror. It's bad to have one's myth shaken up like that. Would Samarkand or Cathay or Cipango have suffered the same fate if visited? As soon as I had cleared the outskirts, the broken-metal-and-glass outer ring, and moved through Mapleton I found a pleasant place to stop on the Maple River not far from Alice—what a wonderful name for a town, Alice. It had 162 inhabitants in 1950 and 124 at the last census—and so much for the population explosion at Alice. Anyway, on the Maple River I drew into a little copse, of sycamore I think, that overhung the stream, and paused to lick my mythological wounds. And I found with joy that the fact of Fargo had in no way disturbed my mind's picture of it. I could still think of Fargo as I always had—blizzard-driven, heat-blasted, dust-raddled. I am happy to report that in the war between reality and romance, reality is not the stronger.

* * *

Someone must have told me about the Missouri River at Bismarck, North Dakota, or I must have read about it. In either case, I hadn't paid attention. I came on it in amazement. Here is where the map should fold. Here is the boundary between east and west. On the Bismarck side it is eastern landscape, eastern grass, with the look and smell of eastern America. Across the Missouri on the Mandan side, it is pure west, with brown grass and water scorings and small outcrops. The two sides of the river might well be a thousand miles apart. As I was not prepared for the Missouri boundary, so I was not prepared for the Bad Lands. They deserve this name. They are like the work of an evil child. Such a place the Fallen Angels might have built as a spite to Heaven, dry and sharp, desolate and dangerous, and for me filled with foreboding. A sense comes from it that it does not like or welcome humans. But humans being what they are, and I being human, I turned off the highway on a shaley road and headed in among the buttes, but with a shyness as though I crashed a party. The road surface tore viciously at my tires and made Rocinante's overloaded springs cry with anguish. What a place for a colony of troglodytes, or better, of trolls. And here's and odd thing. Just as I felt unwanted in this land, so do I feel a reluctance in writing about it.

Presently I saw a man leaning on a two-strand barbed-wire fence, the wires fixed not to posts but to crooked tree limbs stuck in the ground. The man wore a dark hat, and jeans and long jacket washed palest blue with lighter places at knees and elbows. His pale eyes were frosted with sun glare and his lips scaly as snakeskin. A .22 rifle leaned against the fence beside him and on the ground lay a little heap of fur and feathers—rabbits and small birds. I pulled up to speak to him, saw his eyes wash over Rocinante, sweep up the details, and then retire into their sockets. And I found I had nothing to say to him. The "Looks like an early winter," or "Any good fishing hereabouts?" didn't seem to apply. As so we simply brooded at each other.

"Afternoon!"

"Yes, sir," he said.

"Any place nearby where I can buy some eggs?"

"Not real close by 'less you want to go as far as Galva or up to Beach."

"I was set for some scratch-hen eggs."

"Powdered," he said. "My Mrs. gets powdered."

"Lived here long?"

"Yep."

I waited for him to ask something or to say something so we could go on, but he didn't. And as the silence continued, it became more and more impossible to think of something to say. I made one more try. "Does it get very cold here winters?"

"Fairly."

"You talk too much."

He grinned. "That's what my Mrs. says".

"So long," I said, and put the car in gear and moved along. And in my rear-view mirror I couldn't see that he looked after me. He may not be a typical Badlander, but he's one of the few I caught.

A little farther along I stopped at a small house, a section of war-surplus barracks, it looked, but painted white with yellow trim, and with the dying vestiges of a garden, frosted-down germaniums and a few clusters of chrysanthemums, little button things yellow and red-brown. I walked up the path with the certainty that I was being regarded from behind the white window curtains. An old woman answered my knock and gave me the drink of water I asked for and nearly talked my arm off. She was hungry to talk, frantic to talk, about her relatives, her friends, and how she wasn't used to this. For she was not a native and she didn't rightly belong here. Her native clime was a land of milk and honey and had its share of apes and ivory and peacocks. Her voice rattled on as though she was terrified of the silence that would settle when I was gone. As she talked it came to me that she was afraid of this place and, further, that so was I. I felt I wouldn't like to have the night catch me here.

I went into a state of flight, running to get away from the unearthly landscape. And then the late afternoon changed everything. As the sun angled, the buttes and coulees, the cliffs and sculptured hills and ravines lost their burned and dreadful look and glowed with yellow and rich browns and a hundred variations of red and silver gray, all picked out by streaks of coal black. It was so beautiful that I stopped near a thicket of dwarfed and wind-warped cedars and junipers, and once stopped I was caught, trapped in color and dazzled by the clarity of the light. Against the descending sun the battlements were dark and clean-lined, while to the east, where the uninhibited light poured slantwise, the strange landscape shouted with color. And the night, far from being frightful, was lovely beyond thought, for the stars were close, and although there was no moon the starlight made a silver glow in the sky. The air cut the nostrils with dry frost. And for pure pleasure I collected a pile of dry dead cedar branches and built a small fire just to smell the perfume of the burning wood and to hear the excited crackle of the branches. My fire made a dome of yellow light over me, and nearby I heard a screech owl hunting and a barking of coyotes, not howling but the short chuckling bark of the dark of the moon. This is one of the few places I have ever seen where the night was friendlier than the day. And I can easily see how people are driven back to the Bad Lands.

Before I slept I spread a map on my bed, a Charley-tromped map. Beach was not far away, and that would be the end of North Dakota. And coming up would be Montana, where I had never been. That night was so cold that I put on my insulated underwear for pajamas, and when Charley had done his duties and had his biscuits and consumed his usual gallon of water and finally curled up in his place under the bed, I dug out an extra blanket and covered him—all except the tip of his nose—and he sighed and wriggled and gave a great groan of pure ecstatic comfort. And I thought how every safe generality I gathered in my travels was canceled by another. In the night the Bad Lands had become Good Lands. I can't explain it. That's how it was.

Winter Interlude

During the summers of 1983 and 1985, Dayton Duncan retraced the Lewis and Clark expedition's route to the Pacific. Notes on the two summers' travels are merged into one narrative in Duncan's book, Out West: American Journey Along the Lewis and Clark Trail. The next two excerpts describe his experiences along the Lewis and Clark trail in North Dakota. "Winter Interlude" tells of his return to North Dakota for "two frigid weeks" in February, 1985. "Buffalo Hunt" took place the previous summer.

By the time Lewis and Clark were north of what is now Bismarck, they had to begin preparing for winter. It was only late October, but the weather was turning cold, and already there were occasional snows. Their plan was to winter among the Mandan and Hidatsa Indians near the mouth of the Knife River, near present-day Stanton, North Dakota. Here they would build their winter quarters, Fort Mandan, and await better traveling weather. Ice began appearing in the Missouri River on November 13, 1804, and the river was frozen solid by the end of the month; they would not be able to depart until early April 1805.

During the six winter months the explorers spent with these tribes they pursued a number of objectives: preparing reports for President Jefferson on their expedition so far, studying the customs of the village-dwelling Mandans and Hidatsas, seeking as much information as possible on the route they would be taking in the spring, trying to forge a peace between the Arikara tribes to the south and their winter neighbors, and promoting the notion among the tribes that they should trade with the representatives of their new "White Father" in Washington rather than with the English traders from Canada.

Of course, other necessities of survival also took much of their time: finding food (including several buffalo hunts), repairing and making equipment for the next year's travel; being neighborly with the Indians (venereal disease outdistanced frostbite as one of the chief medical problems by midwinter); and just plain struggling to keep warm. The swing between weather extremes in the High Plains is unknown to other regions of America-temperatures top 100 degrees in the summer and crack 40 degrees below zero in winter. References to the cold weather seep through the journals like the frost through the Dakota sod.

Nearly two hundred years later, one thing that time has not changed is the weather. Winters are still brutal in North Dakota.

I probably could have established this simply by checking the national weather maps in the newspaper while sitting by a warm woodstove at home in New England. But a desire, perverse perhaps, to experience a High Plains winter as Lewis and Clark did brings me to Bismarck in early February. Gerard Baker, a young Hidatsa Indian, has arranged for us to spend a few nights in an earth lodge, a reconstruction of the kind of dwelling his people used when the Corps of Discovery was among them.

I arrive in Bismarck by airplane, and the checkout attendant at the car-rental stand gives me the first lesson in the difference between intellectual conception and experience when she hands me a thick coil of wire with the car keys. "Plug it into the car engine and an outlet at night so your car will start in the mornings," she says.

A bank thermometer in downtown Bismarck pegs the unofficial temperature at 19 below at 9 P.M. With the frigid weather is a clear night sky. Thirteen miles from Bismarck, I can still see the city's lights across the rolling plains in the rearview mirror; ahead, to the north, an aurora borealis shimmers on the horizon. The few cars I meet on the road have cardboard sheets across their grilles with a single, small hole punched in the center—this is to allow a little of the winter air into the car's radiator, but not too much. It's best for people to breathe the same way, in short gasps instead of a full inhalation that could seize your engine with pneumonia.

Like people everywhere who live where winters last half the year, North Dakotans try to turn adversity into virtue. Unable to change the weather, or escape from it, they brag about it. "This isn't so bad," the motel clerk tells me when he sees I'm from out of state. "You should have been here last week. It was forty below."

Stanton, North Dakota (population 623), is a town that time has tried to pass by. When river traffic carried people and heavy cargo, the town's location near the confluence of the Knife and Missouri rivers gave Stanton an importance that ended abruptly when the first railroad bridge was built across the Missouri downstream at Bismarck in the last decade of the 1800s. Stanton is still the county seat of Mercer County, but Hazen, about twelve miles west, has a billboard proclaiming itself the largest city in the county, thanks to the boom in the coal business. Even farther west, billboards on the outskirts of Beulah advertise that town as "Lignite Capital of the World." Alternate Route 200, the main highway in these parts, is pulled northwest toward Hazen, missing Stanton entirely. Stanton's a two-bar town with far fewer people living here now than there were in the immediate area when Lewis and Clark encamped for the winter among the nearly four thousand

Mandans and Hidatsas.

Just south of town, a highway sign announces a "scenic turnoff." From the parking space you look down at a bend in the Missouri, an actual river here in the traces below the Garrison Dam upstream and still a number of miles above the outer reaches of the massive Lake Oahe. To the south and east is the smokestack of a coal-burning electric plant. To the north stand two more plants, one on the site of a Mandan village visited by the explorers, surrounded by hills of lignite coal. Just west of these, the tops of giant cranes can be seen as they drag out coal from the bottom of an open pit and place it on a long conveyor belt for delivery to the plants. Electricity is one of North Dakota's chief exports (the state uses only about 40 percent of what it generates). If this is a "scenic turnoff," it must be using slang from the 1960s: it turns me "off," not "on."

Near the parking space is a huge stone, smooth and oval like a pebble you might pick up on the riverbank and fondle in your hand before tossing it into the water, except that this one is about six feet high and twelve feet long. There is no official sign to explain the stone, just the random graffiti from some passerby who probably found the scenic wonders of the coal plants uninteresting and happened to have a few cans of spray paint along to ease the boredom. This stone used to rest on the western bluffs above the river before the highway department moved it down and across the road to anchor the scenic turnoff.

To the Mandan Indians, it was a sacred shrine, intertwined with their legends about the river and the beginnings and ends of things. In this legend, one of several giant turtles that had come from the river to protect and advise the Mandans was convinced not to return to the Missouri. The turtle became the rock, in which state it would remain until just before the end of the world. When the Indian Armageddon arrives, the Mandans are to gather at Turtle Rock and sing on its top. The rock will turn back into a turtle and once again protect the people and advise the survivors. One of the signals that the End is near was to be a split in the rock. The giant stone survived its transfer from the bluffs to the parking lot, but sometime in 1984 two fissures inexplicably appeared that opened its top like a loaf of warm bread. The graffiti survived intact.

When Lewis and Clark reached this area in 1804, the Mandans were already in decline. Continuous wars with other tribes, particularly the Sioux, and a devastating smallpox epidemic in 1782 had seriously reduced their numbers, and, accordingly, weakened the tribe's influence in the Upper Plains. Before the beginning of the nineteenth century, they had moved up the Missouri to live in close and generally friendly proximity to the Hidatsas near the mouth of the Knife River. Like the Hidatsas, the Mandans were village dwellers, not nomads; they lived in semipermanent earth lodges instead of tepees and were noted primarily for their skill in farming rather than following the buffalo herds, although the buffalo was as integral a part of their lives as it was for every other tribe in the West.

With the Hidatsas, the Mandans were the middlemen of the complex trade network of the Upper Missouri. James Ronda, a scholar of Lewis and Clark's relations with the Indians, has called the villages the "great Missouri River country store." They provided not only the produce, principally corn, for other tribes, but also the location for goods to be exchanged: Cheyenne horses and leather products, Assiniboin and Crow furs; goods from Arapaho and Cree, English-made guns and manufactured items, even buffalo meat from their enemies the Sioux were part of their trade.

In 1837, one of the things they got as part of their trade with the whites was another round of smallpox. Because they lived in villages rather than in scattered camps across the Plains, the disease was particularly deadly to the Mandans. It is also said that the hostile Sioux took advantage of the siutation, camping a war party in the area to discourage Mandans and Hidatsas from venturing away while the pox took its toll. In one season, the Mandans were reduced from 1,500 to 150; the Hidatsas went from about 2,000 to 500. Following this catastrophe, the Mandans moved upriver again with the Hidatsas, where they were later joined by the Arikaras.

The Mandans and Hidatsas were unique among the Plains Indians in their steadfast loyalty to the United States government. They never fought the whites. Yet their reward was the same: constant treaty violations and neglect from the government they had befriended and had grown to depend upon. By the mid-1800s they often could not hunt buffalo because the military wouldn't protect them from the increasingly belligerent Sioux. "When we listen to the whites," one chief said in 1870, "we have to sit in our villages, listen to [the Sioux] insults, and have our young men killed and our horses stolen, within sight of our lodges." The women didn't even dare pick wild berries outside the villages.

By the 1880s they had moved upstream again, forming communities along the rich bottomland of the Missouri in northwestern North Dakota. Although their reservation was constantly reduced in size by the government, they began once again to prosper and grow on their ranches and farms. Theirs was a self-sufficient society, not a cash economy. The Depression of the 1930s had little impact on their people, and in the 1940s and early 1950s, only 6 percent of the tribes' members were on public welfare.

But in 1952 the Garrison Dam was completed, and the bottomland, villages, homes, farms, schools, churches, burial grounds—the social and economic base for 85 percent of the people of the Fort Berthold Reservation—gradually disappeared under the new Lake Sakakawea. The three tribes, scattered now on the far less productive high rangeland in five distinct segments of the reservation that sprawls around the serpentine lake, are still struggling to recover. Unemployment and alcoholism, the twin scourges of modern reservations, rose with each foot of lake water.

Nothing much is left of the Mandans as a distinct tribe. The Missouri River, before it was tamed, claimed one village visited by Lewis and Clark; the coal plant sits atop another; and the dam took care of their later villages. Intermarriage with their neighbors the Hidatsas and Arikaras after the smallpox decimated their numbers has resulted in what scholars say are no more full-blooded Mandans, although some people on the reservation claim otherwise.

Just north of Stanton, however, is the Knife River Indian Villages National Historic Site, which encompasses the three Hidatsa villages that were inhabited when the explorers wintered in the area. Huge, circular depressions about forty feet in diameter mark where the earth lodges once stood in the Golden Era of these people. Next to the visitor center is a half-size replica of an earth lodge, built by Gerard Baker in 1979. It will be our weekend home.

At the age of thirty-one, Gerard is an interesting blend of white and red cultures. He has steadily worked his way up the ladder of the National Park Service from an interpreter of native culture to one of the few Indian district rangers with management and supervisory responsibilities. He has a college degree in criminology, which, combined with his size, his even temperament, and his proficiency with weapons, has made him part of the Special Events Team, the Park Service's law-enforcement equivalent of SWAT. Gerard is constantly studying and practicing some of the old Indian ways, partly as an intellectual inquiry into his own roots but mainly as a conscious decision to lead his life by gleaning what he thinks is best from the two, often contradictory, societies. His sources are books, long talks with older people on the Fort Berthold Reservation, and simply trying things out on his own—like building an earth lodge.

Extended conversations with him can range from quotations from the journals of traders at the fur posts in the 1830s to stories of Indian spirits, from hell-raising tales of his days as a basketball star in high school and college to deep thoughts on the future of his tribe. A day in his life might include paperwork at his office, a horse ride to check the park's buffalo herd, scraping and tanning some deer or elk hides, then a nighttime refresher course in cardiopulmonary resuscitation to maintain his emergency-medical-technician license or a meeting with fellow deputy sheriffs in McKenzie County.

He is now district ranger at the Theodore Roosevelt Memorial Park North Unit in the Badlands area more than a hundred miles west of Stanton, but many people in Stanton still remember him from his days starting out with the Park Service at Knife River. In addition to his park duties, Gerard had coached some junior-league basketball teams in town, and his friendliness and sense of humor had made him well known to and well liked by the locals. Besides, Gerard Baker tends to stand out in a crowd, even in towns larger than Stanton. He's six foot five inches tall and weighs about 230 pounds, has long, braided locks of black Indian hair, and is as apt to be wearing leggings he made from some hides he tanned as he is his forest-green park uniform. The historic site nearby, the remnants of the largest of the three former villages, is called "Big Hidatsa." The term could as easily be applied to Gerard.

When we hook up in Stanton, he has on elkskin leggings, a deerskin shirt decorated with porcupine quills and a few beads, a red cloth sash for a belt, and a red *voyageur* cap.

"Do those Indian clothes really keep you warm?" I ask him.

"You bet," he says, then picks up his parka and smiles. "But not as good as ol' Eddie Bauer."

Part of the Indian way, according to Gerard, is teasing, especially as a way of disciplining children. In their culture, uncles and aunts help direct their nieces and nephews toward proper behavior by chiding, kidding, and just plain making fun of the kids. In 1846, when Francis Parkman traveled the West to write The Oregon Trail, he remarked with disgust on the way the Plains Indians parents never rebuked or spanked their children. Teasing by relatives was the Indian substitute Parkman apparently never noticed. In a good-natured way it is also the basis of Gerard's sense of humor.

My knowledge of the camping/outdoor life is rather rudimentary, reduced to three basic rules:

1. Turn down all offers to go snipe hunting
2. Check the direction and force of the wind before relieving yourself outside
3. Watch for a twinkle in the eye of anyone offering suspicious-sounding advice

All three serve me well in the weekend with Gerard. He tells me we might have a "sweat" during one afternoon. We will have to cover willow saplings, bent to make a tiny hut; build a fire to heat some rocks; put the rocks in a hole in the hut's center; cover the saplings with tarps and blankets; disrobe and sit inside while we drip water on the heated rocks and wait to start sweating.

"Hidatsa sauna," Gerard explains. "And then when it's so hot we can't stand it anymore, we jump out and dive into a snowbank."

I look skeptical. The forecast calls for high temperatures in the single numbers.

"You'll go first," he adds. "Indian custom."

A full-size earth lodge was large enough for an extended Indian family of five to twenty-five persons, their personal belongings, and even a few prize horses. The half-size replica Gerard built at Knife River looks like a sod igloo. We have to crawl through the extended entryway to get inside. Four six-foot cottonwood posts, at the corners of a five-foot square, hold up the center. About five feet from them is a ring of smaller posts, between three and four feet tall. Laid across the beams which the posts hold up are cottonwood saplings, then willow branches, prairie grass, and finally a foot of earth. Most of the smoke from the fire we build in the lodge's center escapes through a two-foot-diameter hole at the top. Most, but not all. Enough smoke lingers inside to sting our eyes when we stand near the center.

"Keeps the mosquitoes out," Gerard says, admiring his work while a midwinter snowstorm blows outside.

We spend the gray afternoon getting settled in the lodge. Erik Holland, an archaeologist from Fort Clark and a friend of Gerard's, is with us. We cover the opening of the entryway with a tarp to keep the snowdrifts out and drape a second tarp across the inner opening for good measure. A stack of cottonwood, sap willow, and ash logs rests in one corner of the dirt floor. Keeping the fire roaring for several hours has warmed the interior enough so that we can sit with our winter parkas off, even though the mercury outside is already below zero and still falling.

Gerard takes out a small twined rope of sweetgrass, which he ignites in the fire and then blows out, smudging the lodge in each of the four directions, and finally places the remainder over the southern entrance.

"For the spirits," he says. His ancestors are buried nearby.

It's getting dark now, and the driving snow has abated, replaced by a driving wind under a cloudless slate sky. Gerard puts a large pan on one edge of the fire and begins boiling potatoes, onions, some red peppers, and buffalo tripe. Rather than the mere stomach lining of the buffalo, this is not just tripe, it's the buffalo "book," thin leaves of membrane that assist the beast's digestion and look like a medium-sized encyclopedia unfolding in the boiling water.

Gerard considers it a delicacy. "Hidatsa lobster," he calls it, in honor of my New England home. He scoops out a bowl of the concoction and hands it to me across the yellow cone of firelight.

"You go first," he says, and watches intently as I take a few tentative bites. I catch a twinkle in his eyes before he turns to Erik and says, "I guess it's not too hot for us to eat," and fishes out two more servings. I wouldn't rate buffalo book with boiled lobster, but the hunger that accompanies a North Dakota winter day only adds to the adage that beggars can't be choosers. We eat the greasy, rubbery meal with our hands; it makes more sense than chasing it around our plates and the earth floor with forks and knives.

Gerard has brought five buffalo robes with him to the earth lodge. He tells me to place one on the ground, fur up, and use the other four, fur down, over me for the night. Before I can ask him how he plans to keep warm while he sleeps, he starts unrolling his down-filled sleeping bag near the fire, opposite the entryway. Erik Holland has one as well.

When I inquire about the relative warmth of buffalo hides and sleeping bags, Gerard shrugs and says the hides served the buffalo and the Indians pretty well in the winters of the past. "I do know my sleeping bag's guaranteed to twenty below," he adds.

He then starts concocting Indian traditions for my benefit. For instance, he says whoever has the most buffalo robes in an earth lodge is responsible for keeping the fire going throughout the night.

After we lie down to sleep in our respective places, there is a long silence; then I hear his voice, muffled through his sleeping bag.

"Can you tell the difference between sap willow and cottonwood by their feel in the dark?"

"No," I answer. "Why?"

"Just curious. They'll both burn okay, but sap willow pops in a fire. A spark landing in the wrong place can play hell with an old buffalo robe. Good night."

Fully clothed in my long underwear, insulated socks, jeans, shirt, sweater, and stocking cap, my rolled-up parka as a pillow and the oppressive weight of four buffalo robes on top of me, I go to sleep, listening to the rush of the wind across the smoke hole. The wind's wail is interrupted by the occasional pop of a sapwillow log.

About two and a half hours into the night, something wakes me up. Coldness. Not a chill. Not some little shiver that comes when you shift your feet under the blankets at night and the move from the envelope of body warmth makes you grunt and toss in your sleep. Coldness—as if frost is trying to work its way into you from the top of your head and the bottoms of your feet. Painful coldness.

The fire has burned down to embers. It gives off as much heat as it does light, a mere memory of both. I lie awake and wait, hoping either Gerard or Erik will wake up and feed the fire. Nothing happens. The frost line creeps another tenth of an inch into my body. Gerard was right: the person with the most buffalo robes is the one who tends the fire, if everyone else is in goose down and nylon. Buffalo robes don't carry the same guarantee. I shed the robes and feel my way to the woodpile, drag some logs to the fire, and get back under the buffalo hides. The flame builds up—with a few pops from some sap willow—and creates just enough heat to let me go back to sleep. The same routine repeats itself twice more during the night.

Cold feet, even colder than during the night, wake me early in the morning.

I have the choice of closing my eyes, lying still, and failing to get back to sleep because my feet won't let me, or staying awake while I move them back and forth under the hides to warm them up. Eyes shut or open, moving my feet or not moving, it's still not sleeping. At least the second option might ward off frostbite. Gerard and Erik are still asleep. I can tell this by the slow, rhythmic puffs of breath that rise from one end of their sleeping bags, smoke signals to me that their feet are warm.

As I lie on my back, sort of jogging in place underneath the buffalo robes, I watch the morning arrive through the earth lodge's smoke hole. A box elder tree hovers over the opening. The changing contrast of colors between the tree branches and the sky marks the passage of time to daybreak: black sky with even blacker branches gives way to dark blue, then silver etches against a slate background, then tawny lines across gray, and finally brown tree branches in the robin's-egg blue of day. From nightmarelike Rorschach test, to Impressionist painting, to some sort of nature color photograph through a peephole, the scene unfolds, like the awakening of conscious thought in a human being—the changes imperceptible as they occur and made known only by the uneasy sense of loss of what preceded. Through it all, the box elder twitches and shudders in the blasts of wind.

There's enough light inside to look around the lodge. The top inch of water in our plastic bucket is ice, even though it was only four feet from the fire. I can see now the reason for my frozen feet: a small drift of snow angles in from the entryway, coils behind the woodpile, and stops in a heap next to the bottom of my buffalo hides. Someone must have been sloppy in readjusting the tarp when he returned from answering a call of nature in the night. It's something that can be easily understood and forgiven if you've ever tried to unwrap yourself from a warm cocoon, put on boots, crawl out of an earth lodge into twenty-below darkness, do your business in 30 mph winds, and retrace the whole thing in the space of 2.3 seconds or less.

My ballpoint pen is too frozen to work. The cold has sapped the life out of my tape recorder's batteries. So I busy myself building the fire into an inferno and pondering the winter spent by the explorers. Their journals offer at least the company of cold misery. "Sent out 7 men to hunt for the Buffalow they found the weather too cold & returned," Clark writes in mid-December. Temperatures of 20 below zero, "74 degrees below the freesing pointe," minus 21, minus 40 are noted. One entry is particularly intriguing and relevant to the night I've just experienced: "this day being Cold Several men returned a little frost bit, one of [the] men with his feet badly frost bit my Servents feet also frosted & his P—s a little." I don't feel so bad in comparison.

I will always remember Erik Holland fondly. Gerard and I are up and talking about boiling coffee grounds in the unwashed tripe pan, when Erik's voice interrupts us from his sleeping bag.

"I'll buy breakfast for everyone at the Riverside," he says.

Less than fifteen minutes later, we're in Stanton ordering ham and eggs, drinking good black coffee, and the two of them are telling me about waking up in the middle of the night and waiting silently for someone else to put more wood on the fire.

"One of you could have done that miserable work," I complain.

"You go first," they say in unison.

At least I wasn't the one who left the tarp half open, I tell them.

"How do you know?" Gerard asks.

"I must have gone first."

The chief topics of conversation among the other patrons at the Riverside are the basketball team and the weather. Both have fallen on cold times, and there seems to be nothing anyone can do about either, except talk about them. The Stanton High School teams are nicknamed the Sioux, a particular irony. Imagine the reactions of the spirits of the Hidatsa and Mandan warriors, buried less than a mile away, when they hear the yells from the gymnasium each Friday night: "Let's go Sioux!" No wonder the team seems jinxed.

Breakfast finished, we head down Alternate Route 200 so we can cross the river and visit the reconstruction of Fort Mandan, Lewis and Clark's winter quarters. Erik is driving, I'm in back. Gerard, in the shotgun seat, is playing some ceremonial Hidatsa chants and songs on the tape deck. "Hidatsa Top Ten," he shouts back at me over the blare and returns to singing along. Each song is different, he tells me, which I have to accept on faith. They're all indecipherable to my ears—just a series of "*hey*-yay-yay-yah," "*hey*-yay-yay-yahs" at various pitches—and sound the same to me. Now I know how my parents felt about the rock and roll I played to them as a teenager.

We stop briefly at Turtle Rock, where Gerard explains the Mandan legend and disdainfully inspects the graffiti. Erik points out the location of the principal Mandan village during the explorers' time, now under a power plant and coal slagheap. The car is still running, and the tape of chants is audible from where we're standing by the rock. Off in the distance, the other coal plants send their smoke signals pluming hundreds of feet into the cold blue sky, communicating the new order of the universe. Conical piles of strip-mined coal sit like menacing Sioux tepees on the bluffs overlooking Big Hidatsa.

The sun is now high in the sky, with no clouds in sight except those created by the smokestacks, yet the temperature is still two below zero. It's as if the sun had given up trying to warm North Dakota. An incessant wind blows thin lines of snow across the highway, spirits on a hurried journey to join compatriots in the drifts wherever the ground dips.

The fissure in Turtle Rock is supposed to signal the Coming of the End. Maybe it's already arrived, not in some sudden cataclysm, but in the passing of a people and their way of life. Maybe the Turtle waited in vain for nonexistent Mandans to gather on its back. Freed from its stony prison and without any people to protect, perhaps the Turtle slipped silently back into the Missouri, where it now plies the waters between two symbols of the greater Spirit that defeated it, the Oahe and Garrison dams.

A local historical group has built a replica of Fort Mandan among the cottonwoods on the east bank of the river. The actual location is a little ways upstream and under water, a victim of the earlier, wandering days of the Missouri.

"You go first," Gerard tells me as we trudge through the town toward the palisades of the fort, "in case someone's still there. They might get the wrong idea if they see an Hidatsa in buckskins."

This gives him the idea for my Indian name. According to Indian tradition, a person is given a name by his elders, usually in conjunction with some event in the person's life or some personal characteristic.

"Man Who Sleeps in Buffalo Robes" or "Buffalo Robe" was his first idea last night. Then "Smells Like Tripe" became a favorite. But now he settles on "He Goes First"—*Ee zeek' ah nets*, it sounds like in Hidatsa—and it sticks.

The fort is triangular, a design Clark learned during his military campaigns in the Ohio River valley before the expedition. Two sides of the triangle form living quarters, the outside walls tall and the roofs sloping toward the center courtyard; the third wall is a twenty-foot palisade with a gate. Compared to our half-size earth lodge, it seems commodious, though cold and barren sitting uninhabited. Filled with a company of forty-five busy men, it was no doubt crowded and at least as warm as we were last night.

In something like this the explorers had spent a long, frigid North Dakota winter, becoming friendly with the large villages just up-stream, waiting for spring, and preparing to venture into territory no whites had crossed before them.

"He Goes First." I like the name, despite the jokes about my own experiences it implies. It makes me feel closer to Lewis and Clark than I've felt so far in their footsteps.

New Town, North Dakota (population 1,335), the Fort Berthold Reservation agency town, was created when Lake Sakakawea inundated the river towns of Beaver Creek, Nishu, Charging Eagle, Lucky Mound, Shell Creek, Independence, and Elbowoods.

New Town is a four-bar town without a stoplight. The local movie theater, the Trail, is open only Thursdays through Sundays, but videos are available every day at the Rexall drugstore. A gas station offers a free video rental with every ten gallons of gas. Two evenings in a row, I lose money at the two-dollar blackjack table at the Steak 'n' Stein. The Sunset Motel, where I pass a few nights, is inexplicably on the east side of town. Gerard has warned me away from two of the bars in town unless I am willing to fight my way out. There was no twinkle in his eye when he gave me this advice, and so I follow it.

Throughout the reservation the housing—much of it government-built—consists of small ranch-style homes on barren sites. They look more solitary and exposed than the rounded humps of a cluster of earth lodges must have appeared when Lewis and Clark visited the area.

Alyce Spotted Bear, the chairwoman of the Three Affiliated Tribes (Mandan, Hidatsa, and Arikara), brings me up to date on the tribes' history. Besides flooding their bottomland farms and unalterably ending the tribes' self-sufficient existence, the Garrison Dam and Lake Sakakawea dispersed the tribal members onto upland sections of the reservation separated from each other by drives of two to four hours, she says. In her view, the sociological havoc wreaked on their communal spirit by the dam since the 1950s has been as devastating as the smallpox epidemic of 1837.

Her two main hopes for economic recovery are employment at a giant coal gasification plant near Beulah and fostering tourism—including a casino—on tribal lands near the lake.

A resort hotel is already complete on a treeless plain west of New Town. It is named Four Bears Motor Lodge, after the great Mandan chief. Before going to sleep one night there, I take out one of the books about tribal history that Gerard has loaned me and happen upon the last speech Four Bears ever gave. He was on his deathbed from smallpox in 1837:

> Ever since I can remember, I have loved the whites. I have lived with them ever since I was a boy and to the best of my knowledge, I have never wronged a white man. On the contrary, I have always protected them from the insults of others, which they cannot deny. The Four Bears never saw a white man hungry, but what he gave them to eat, drink and a buffalo skin to sleep on in time of need. I was always ready to die for them, which they cannot deny. I have done every-thing that a red skin could do for them, and how they have repaid it! With ingratitude! I have never called a white man a dog, but today I do pronounce them to be a set of black-hearted dogs. They have deceived me. Them that I always considered as brothers have turned out to be my worst enemies.
>
> I have been in many battles, and often wounded, but the wounds of my enemies I exult in. But today I am wounded, and by whom, by those same white dogs I have always considered and treated as brothers.
>
> I do not fear death, my friends. You know it. But to die with my

face rotten, that even the wolves will shrink with horror at seeing me, and say to themselves, "That is the Four Bears, the friend of the whites."

Listen well to what I have to say, as it will be the last time you will hear me. Think of your wives, children, brothers, sisters, friends, and in fact all that you hold dear. All are dead, or dying, with their faces all rotten, caused by those dogs the whites. Think of all that my friends, and rise together and not leave one of them alive.

Before turning out the ligbt for the night, as a white tourist at the resort lodge named for Four Bears, I get up from the bed to make sure the door is locked.

Gerard introduces me to some of the older tribal members, who describe life before the dam. Each time, as one talks of the bottomlands—men often mention good winter shelter for their cattle and horses, women the wild berries and fruits of summer—the dark eyes of the old person glisten with mist.

Gerard's mother feeds us a meal in her country home that reminds me of dinner at my grandmother's: fresh-baked rolls, heaping slabs of roast beef, plenty of vegetables, and a cake for dessert. Mrs. Baker stands near the table and watches approvingly as Gerard and I, his wife, Mary Kay, and their three children devour the food she has prepared. According to older customs, she will eat only after everyone else is done.

Paige Baker, Sr., Gerard's father, was a member of the reservation council, a tribal judge, and on a number of North Dakota boards and commissions. He and his wife, Cora, insisted that their four children get a good education. Fred Baker, their oldest child, is now an administrator for the Public Health Service at the Rosebud Reservation in South Dakota; Paige Jr. is completing a doctorate in administration at Penn State University; and Mary is dean of academic affairs at the United Tribes college in Bismarck.

Their home, where the Bakers relocated when the lake flooded Independence in 1952, a year before Gerard was born, sits in a wooded draw on the western edge of the reservation. The small living room is filled with the iconography of a close-knit family: graduation pictures of the children, photos of the grandchildren, childhood mementoes, a Catholic crucifix. One photograph shows Paige Jr., dressed in traditional Hidatsa clothing, meeting Pope Paul; the cover of an outdated North Dakota state map features Mary in a colorfully beaded buckskin dress. (A Baker relative served as the model for the Sacagawea statue that stands near the state capitol.)

Gerard loans me a tape cassette of some conversations with his father, made before Paige Baker, Sr., died of lung cancer. The tapes are filled equally with reminiscences and fatherly advice. At one point, Mr. Baker tells of his grandmother taking him to an ash tree in 1921 and crying that the spiritual things of the tribe were being lost; then, with her grandson next to her, she implored the tree to give her grandchildren some of its strength to survive.

"Now everything is lost," Mr. Baker says on the tape. "We don't believe in the white man's way and we don't believe in the Indian way. So what the hell are we? We're not anybody." He coughs several times, and it's easy to imagine him collecting his thoughts before turning to his youngest son with words to remember once he's gone: "You're going to run into trouble; you're going to look for something. You'll wish you had a god or something more powerful than you are to pray to and then hang on to, which you're not going to have if you don't practice it now and learn about it. But if you learn about it, then when the time comes that you need that help, you'll know you've got it. Hang on to things like that."

When Paige Baker, Sr., died, Gerard tells me, their small herd of horses arrayed themselves in a straight line facing the road when the hearse with his father's body passed by on its way to the cemetery. The elder Bakers were practicing Catholics, but knowledgeable and respectful of the Hidatsa ways; Gerard reversed the order.

He started learning the "old ways" and began believing. He read books about his heritage—the Lewis and Clark journals, diaries of fur-company traders, Prince Maximilian's writings, anthropologists' studies. He looked at Bodmer's drawings. He started collecting oral histories from older tribal members. He skinned animals and tanned their hides, made traditional clothing, built tepees and earth lodges and sweat lodges, became an expert marksman with a black-powder musket.

He spent long times outdoors, fasting for several days and going off alone into the eerie buttes of the Badlands to wait for spirits to approach him. The first one was a man coughing behind him as he sat on a hillside at night. While he was singing Hidatsa songs in a coulee one day, a woman's voice started singing with him. Birds spoke to him in his dreams.

His life became not a rejection of the white man's way or the Indian way, but a unique synthesis of both. One morning, after he has told me about his belief in spirits, he goes outside to the basketball hoop in the driveway to teach his nine-year-old son Bobby a hook shot.

Like the Mandans and Hidatsas with whom Lewis and Clark spent their North Dakota winter, Gerard is generous with his hospitality toward me. And like the explorers, who gleaned information about their hosts through persistent questioning, I compile Gerard's life story not from his offering it up—self-promotion or bragging is unseemly to the ways of the Hidatsa—but through constant probing on my part. His hospitality includes answering a guest's questions honestly and directly, regardless of how many or how personal they may be.

"Everything on this earth, including trees, including rocks even, everything that's here has a spirit, and therefore we should respect it," he says during one conversation. But the spirit I grow to respect the most during these two weeks is his.

On my last day in February with Gerard, we drive along a winding road on the high bluffs overlooking the Little Missouri River. When we get out of his truck, there is a pungent sulfurous smell in the air from escaped gas at the wellheads of the oil rigs ringing the park. Three buffalo bulls are grazing about a quarter mile from us, and we walk a hundred yards in their direction so I can get a picture of them.

"You can earn an eagle feather by pulling one of their tails," he jokes.

I raise my camera and look through its long lens. We are upwind from the threesome. Smelling us, they lift their massive heads and begin trotting in our direction.

"They seem to be headed our way, Gerard," I say, standing my ground to snap the pictures. "Gerard—?"

When I turn around he is already halfway to the truck at a dead run, motioning me to follow.

"You can earn your feather in the summer," he tells me back at the truck. "We'll take a sweat and then go see the buffalo."

During the airplane ride back to New England, people near me seem to wrinkle their noses. I am a different kind of synthesis of white and Indian ways. My parka still reeks of earth-lodge smoke and buffalo tripe.

But I wear the scent, like my new name, with a certain pride. When we arrive at the airport gate and it's time to walk off the plane, I go first.

Buffalo Hunt

The highways that take me from Mobridge, South Dakota, to the north unit of the Theodore Roosevelt National Park in western North Dakota must have been easy work for the crews that painted the yellow center lines. U.S. 12 west, North Dakota Highway 22 north, Highway 200 west, and U.S. 85 north rarely bend. The painters could have locked the steering wheel of their truck, set the paint machine on automatic, and taken long naps across fifty-mile stretches while their job got done.

Within a half hour from the end of the Memorial Day service at Sitting Bull's grave, the Discovery has taken over the task of travel, allowing my mind to return to my winter stay with Gerard Baker. Dirty white clouds cloak the May sky—the same color and texture as the snow on the North Dakota plains in February.

Memories of the cold weather, the gold plume of smoke rising through an earth lodge's center hole into the winter night, the walk through deep snow to Fort Mandan, the buffalo that chased us back to our truck: those are the mile markers of my thoughts while the camper's odometer clicks off the distance on the straight pavement.

A change in the road snaps me briefly out of the past and into the present. About fifteen miles south of Watford City, North Dakota, U.S. 85 crests a lip of plains and plummets into the Badlands of the Little Missouri, a jumble of eroded coulees and towering buttes whose vertical sides are lined with strata of sandstone, bentonite clay, shale, and veins of lignite coal. The setting spring sun paints the surrounding scene in soft tones of pink, yellow, blue, and purple. A sign welcomes me to the Theodore Roosevelt National Park's north unit. It transports me back in time once more.

Beginning thousands of years ago and continuing into modern times, lightning bolts against the buttes have set the layers of lignite on fire, sending smoke out of fissures in the hillsides and baking the neighboring seams of sand and clay into a red scoria.

General Alfred Sully, chasing the Sioux through the area in 1864, called the landscape "hell with the fires out." Nineteen years later, a young man from the East touring the Badlands for different purposes had the opposite reaction. Theodore Roosevelt had come from New York to hunt buffalo and strengthen his asthmatic lungs in the semiarid air of the High Plains. Within just a few days along the Little Missouri, he decided to start a cattle ranch in the Badlands. "Here," he said later, "the romance of my life began."

For roughly a decade Roosevelt divided his time between his ranch in North Dakota and his interests back East. His months at his Elkhorn ranch on the west bank of the Little Missouri would be spent in cattle roundups, long rides across the broken landscape, hunting expeditions, and even a notorious capture of three Dakota desperadoes (like Gerard, Roosevelt was a deputy sheriff). Back East, he would dine with the intellectual and social elite of the upper classes, write essays and books, and engage in reform politics.

The Badlands cowboys first called the scrawny, bespectacled Easterner "Four Eyes," but his grit and boundless energy soon changed that to a respectful "Mr. Roosevelt." The transfer of influence went both ways. Life on the frontier of the West profoundly shaped the views of the man who would become president. His dual existence worked its own synthesis on Roosevelt and his thinking: which found its expression in his theories of Americanism.

"Like all Americans," he told a Fourth of July rally in Dickinson, North Dakota, in 1886, "I like big things; big prairies, big forests and mountains, big wheat-fields, railroads, and herds of cattle too, big factories, steamboats, and everything else. . . . I am, myself, at heart as much a Westerner as an Easterner."

On his way to the White House, Roosevelt would write a glowing biography of Senator Thomas Hart Benton of Missouri, the champion of Manifest Destiny, and a multivolumed The Winning of the West, which glorifies the "restless and reckless hunters, the hard, dogged, frontier farmers" who pushed the nation's boundaries to the Pacific coast.

Becoming president eleven years after the Census Bureau declared the official close of the frontier in 1890, Roosevelt exercised his Americanism abroad by making the United States a muscular world power, and at home by fostering the creation of national parks and wildlife refuges. In both—his love of wild places and his restless, some would say reckless, expansionism—the influence of his days in the West reverberated for generations.

The long drive from Sitting Bull's grave to the park has cramped my shoulders; the even longer travels in time have turned my mind to mush. I'm ready for a "sweat" when I pull into Gerard Baker's driveway. Within minutes of my arrival Gerard, his son Bobby, and I are traveling to his mother's house, where Gerard already has the willow saplings bent to form a sweat lodge next to the horse corral.

We start a fire with old ash fenceposts and cover the lodge frame with tarps and blankets. Gerard has invited two of his younger cousins to join us. Their admiration for him is clear in the deference they pay him and the way they listen to his stories. He teases them constantly.

The northwest corner of the sky is black with storm clouds. While we heat stones in the open fire, the family's string of horses gallops into the draw, and one of the cousins asks Gerard why they're running.

"Storm coming," he says. "Change in pressure." Three dogs round the hill, nipping at the horses' heels. "That and the dogs," Gerard adds.

Mosquitoes start biting us as we disrobe to enter the sweat lodge. "Storm coming," Gerard says as he slaps one. "Change in pressure," we all answer.

The sweat lodge is too small for all five of us. Gerard and I go first. We shovel some of the red-hot stones from the fire into a small hole at the lodge's center and crawl in, closing the tarps behind us, and sit cross-legged and stooped in the tiny enclosure.

It is dark inside, except for the glow of the rocks, which draws my stare deeper and deeper, much as the intense heat draws my perspiration each time Gerard drips some water on the rocks. We are cleansing "bad spirits" from our body, Gerard says, and then he relates another tale from tribal mythology about the end of the world. The Lone Man, a dominant figure in their lore, told his people the end would be near when four things happened: when the Missouri River (called the Grandfather) ran backward, when the prairie grass turned upside down, when the trees turned to show their roots, and when the field mice grew antlers.

Dams have pushed the river back over the tribes' homeland, farmers have turned the sod, and coal shovels have opened giant pits where trees once stood, Gerard says; all that's left is some sort of mutation of field mice, perhaps by atomic poisoning or genetic experimentation.

The glow of the rocks deepens. Gerard drips on more water, and we sit and sweat in silence until the heat finally forces us out into the night. Bobby and the cousins crawl in to take our place. Gerard and I stand naked in the dark coolness.

"Storm coming," he says again. The wind picks up and big drops of May rain wash our bodies.

My night's reading from the journal includes Clark's entry at Fort Mandan on January 5, 1805:

> A Buffalow Dance (or Medeson) for 3 nights passed in the 1st Village, a curious Custom the old men arrange themselves in a circle & after Smoke[ing] a pipe which is handed them by a young man, Dress[ed] up for the purpose, the young men who have their wives back of the Circle go [each] to one of the old men with a whining tone and request the old man to take his wife (who presents [herself] necked except a robe) and—(or Sleep with her) the Girl then takes the Old Man (who verry often can scarcely walk) and leades him to a convenient place for the business, after which they return to the lodge; if the old man (or a white man) returns to the lodge without gratifying

> the Man & his wife, he offers her again and again; it is often the Case that after the 2d time without Kissing the Husband throws a new robe over the old man &c. and begs him not to dispise him & his wife (We Sent a man to this Medisan Dance last night, they gave him 4 Girls) all this to cause the buffalow to Come near So that they may Kill them.

The buffalo dance was not an exercise in promiscuity for the Mandans of Lewis and Clark's time. It was a religious ceremony whose purpose was to transfer the "medicine" of respected elders or visitors from other cultures to the buffalo hunters. The custom was abandoned by the tribe long ago.

"Did you dream of your girlfriend?" Gerard asks me in the morning. "If you did, it'll help us find the buffalo." He knows I read the journals last night.

Two of the 150 buffalo under Gerard's supervision have escaped the parks' reinforced-fence boundaries. Regulations springing from concern about loose buffalo destroying ranch fences and trampling crops require him to find them. If they're still close enough to the park, he tries to herd them back; if they're too far away, he has to shoot them.

An attendant at park headquarters calls with the news that a rancher spotted the two buffalo southwest of the park early this morning. Gerard puts two hunting knives, a 30.06 rifle, a box of ammunition, and a bow saw in the pickup; Mary Kay hands us a sack lunch as we head out the door. The morning is clear and hot. Last night's rain has scrubbed the buttes of the Badlands, and the wet sage gives off a sharp, sweet smell.

"They're just like us Indians, the buffalo," Gerard says as we drive off. "They've lost everything they used to have. They've dwindled down. They've been put down in confined areas just like we are. They're protected, they're taken care of by Uncle Sam. Back home it was the same way they gave us commodities and canned food to eat. Sometimes the buffalo are a pain in the ass. I imagine sometimes we're a pain in the ass, too. But they're still proud, boy, those old buffalo."

We cruise back roads about five miles from the park for an hour, looking in vain for two dark humps in the tan coulees. The gravel roads are straight. My eyes scan the distances; my mind recedes to the past.

That the buffalo could have gone from more than thirty million to a few hundred and the brink of extinction in the space of a half century is a staggering notion. Never mind the degree of senseless slaughter it implies, it seems a logistical impossibility. But it happened, and not by chance of nature but by design of man.

First came the market forces. Buffalo Bill Cody made a good living providing buffalo meat to the railroad crews and leading Army generals and European aristocrats on extravagant hunts. In one season alone, he boasted of having singlehandedly killed 4,280. Lucrative as the buffalo-meat and buffalo-robe business was, however, the big money was made in the 1870s, after new technology in tanning created increased demand for the buffalo hides. Buffalo leather was used in everything from machinery belting to home furniture. Now the slaughter was unceasing. Fortunes could be made.

During the depression after the panic of 1873, buffalo hunting was one of the few profitable businesses; hunters and skinners swarmed over the Plains. Hunters perfected a new technique: the "stand." Rather than chasing a herd on horseback, a hunter would use the terrain and wind to his advantage, sneak up on a herd, and begin a slow, deliberate shooting of the lead animals. If he was a good enough shot, he would drop the beasts one by one while the others grazed without stampeding. One hunter set a record by killing 120 buffalo during one "stand" of forty minutes. The barrel of his rifle finally overheated from the rapid firing.

The herds thinned. In 1873, only a year after reporting that there was "apparently no limit to the numbers of buffalo," Colonel Richard Irving Dodge found rotting carcasses in their place. "The air was foul with a sickening stench, and the vast plain, which only a short twelvemonth before teemed with animal life, was a dead, solitary, putrid desert," he wrote.

Such reports spurred the House and Senate in 1874 to pass a bill prohibiting the killing of a buffalo cow by anyone except an Indian and making it illegal to kill more animals than could be used for food. But President Ulysses S. Grant, more concerned with hostile Indians than dwindling buffalo herds, stopped the bill with a pocket veto. Government policy would finish what the marketplace started.

When the Texas legislature was poised to pass its own bill protecting the vanishing herds in the state, General Phil Sheridan, Grant's chief agent in handling the Indians, hastened to Austin to address a joint assembly. Sheridan, who coined the phrase "the only good Indian is a dead Indian," persuaded the Texans against the legislation with this argument:

> These men [buffalo hunters] have done in the last two years, and will do in the next year, more to settle the vexed Indian question than the entire regular army has done in the last thirty years. They are destroying the Indians' commissary; and it is a well-known fact that an army losing its base of supplies is placed at great disadvantage. Send them powder and lead, if you will; but for the sake of a lasting peace, let them kill, skin and sell until the buffaloes are exterminated. Then your prairies can be covered with speckled cattle and the festive cowboy, who follows the hunter as the second forerunner of an advanced civilization.

Sheridan suggested that instead of being stopped, the hunters should be given bronze medallions "with a dead buffalo on one side and a discouraged Indian on the other." Years later, with both Indian and buffalo reduced in number and safely confined, a different version of Sheridan's medallion would be circulated by the government in the form of nickel coin. The buffalo on the nickel was alive and standing; the Indian on the other side looked rather noble.

By the time Roosevelt first came to North Dakota in 1883, the hide companies were bankrupt from the virtual disappearance of their source of supply. For several cold and rainy days Roosevelt and a guide scoured the Badlands. They saw only a few buffalo. Teddy fired on one, but his shot hit the animal too far back from its head. As the wounded and angry bull charged, Roosevelt's pony panicked and knocked his rifle against his head; luckily, the buffalo veered off and disappeared. Roosevelt escaped with a bloody face. (Accidents like this were not unusual: in the scramble of a mounted hunt, Custer had once misfired his gun, killing his wife's favorite horse underneath him.)

Characteristically undaunted, Roosevelt continued his hunt. Caught without shelter during a night's rainstorm, lying in blankets in a puddle, the Easterner surprised his guide by turning to say: "By Godfrey, but this is fun!" Later in the month, he finally bagged a bull.

After a survey of the West in 1894 found only eighty-five free-ranging buffalo (about a thousand more were in parks, zoos, and private herds), Congress passed a law protecting the survivors. In 1905, the American Bison Society was formed to encourage the preservation and propagation of buffalo. Theodore Roosevelt, by then president of the United States, was named honorary president of the society.

An excited rancher flags us down from his pickup. He's seen the two bulls and directs Gerard to their location. When we get there, we can barely see them from the gravel road, two black dots a half mile distant and moving steadily northwest along a creek bed. Gerard checks his map and comes up with two pieces of bad news: they're already more than twelve miles from the park perimeter, getting farther with each step; and the closest road in their direction is another fifteen miles away. To follow them, we'll have to take the Dodge pickup into the rough breaks; since they're so far from the park, he'll have no alternative but to shoot them when and if we catch them.

Gerard talks to the park headquarters through his two-way radio, telling them our location and dispatching another crew to start west and head off the buffalo in case our pursuit fails. We open a gate and start across the rangeland. It's noon.

A mile of bouncing between sagebrush and cow pies brings us to a problem. A tributary of the creek the buffalo and we are following enters from the left. We're hemmed in on two sides by seven-foot drops to the stream beds. No problem for the buffalo, who can descend and climb like oversized mountain goats. Big problem for the pickup, even if it had four-wheel drive, which it doesn't. The buffalo are now more than a mile in front of us, walking a steady gait in a straight line, as indifferent to the danger we represent to them as they were to the ditch that has stopped our progress.

We turn ninety degrees to the left of the bulls' path—our only option other than giving up and going back—in hopes of finding a place to cross the tributary. Instead, we find a tributary to the tributary, another deep gully that forces us left again and heads us at an oblique angle to the projected course of the two buffalo. The bulls have long since disappeared behind a broken string of buttes.

Last night's rain has settled the dust of the sandy soil in the valley. But in some of the low places of the faint trace we're following, the earth is deceptively muddy. Several times we nearly get stuck, a prospect both of us dread this deep into the Badlands.

"One of us is going to have to walk in front of the truck to watch for wet spots," Gerard says. It's a park truck; he's the only park employee. A faint twinkle starts to form in his brown eyes, but I beat him to the punch.

"I'll go first," I say and climb out, "but I think in the old days it was the other way around. The Indians did the scouting."

"Did I ever tell you that you can earn an eagle feather by pulling the tail of a rattler?" he shouts from the truck window as I start a half trot ten yards in front.

For about an hour our strange procession loops and winds toward higher ground and a creek crossing, a late twentieth-century miniature wagon train led by someone who's never been in this country before. After several dead ends, false trails, half circles around solitary buttes, climbs and dips, crossings and recrossings of shallow gullies, we emerge on the opposite side of the line of buttes where we last saw our quarry. Just before two o'clock we find a break in the chain, I get back into the truck, and we drive through to the main stream bed. The two bulls are 150 yards in front of us.

The truth is, I had dreamed of my girlfriend last night.

The buffalo seem unconcerned about our presence behind them. They just keep walking at their steady gait-about 5 mph-swishing their long tails as they lumber along. Their sides are mangy and spotty from shedding their winter coats. The smaller, younger bull, about the size of a big ox, occasionally swings his massive head to the side to look at the pickup trailing them. The older bull, half again as large as his young companion, is in the lead and never once looks back.

As if this were their destination all along, and they had been wondering when we'd show up to join them, the bulls round a small hummock and stop next to an untended oil well. The two buffalo begin grazing on tufts of short grass. The pump at the well rocks back andforth, dredging the liquefied remains of the buffalo's prehistoric ancestors hundreds of feet below us.

Gerard uses this pause to check his map. The well is too recent to show up on the map grids marked in ownership sections of green (Forest Service), blue (state land), purple (Park Service), and white (private land). Still, it's easy to estimate our approximate location, despite the wandering path that brought us here. Most of the oil production in this region is on leases from the government; we're on a square green island in a sea of surrounding white sections on the map. The Theodore Roosevelt park is at least twenty miles of broken country to the east, and the buffalo clearly aren't heading back. Once they move from this well site, they'll be walking deeper into private lands and deeper into trouble. A red-scoria access road curls into the pump station. If Gerard is going to have to kill the bulls and have any chance to remove their carcasses, this is where it will have to happen.

He calls the patrol car he dispatched earlier on his radio, then park headquarters for another pickup to haul the second carcass. He radios the sheriffs dispatcher for McKenzie County to see if any deputies are nearby with rifles, in case the second bull takes off at the first shot; the dispatcher tells him it would be more than an hour before a deputy could show up.

The bulls have finished grazing and follow the scoria road for a few hundred feet and then start up the side of a barren hill. We flank them in the truck and stop. Gerard loads the rifle with 150-grain shells, opens the truck door, and draws a bead on the big bull, using the open window as a brace. He mutters something in Hidatsa before he pulls the trigger.

The first shot hits the older bull in the side of the head. He lifts his head and shakes it, as if startled from sleep by a troubling dream.Then he turns his rump toward us, and remains stationary on his feet. The younger bull has jumped, but settled, from the report of the rifle. Gerard says something again in Hidatsa, takes aim, and fires. The young one, which the bullet hits right behind the ear, breaking his neck and killing him instantly, crumples to the ground.

The old bull has turned again now, offering a clear second shot. He is coughing gallons of brilliant red blood onto the sandy soil. More Hidatsa from Gerard, another loud crack from the rifle, and the bull drops like the other one, chin on the ground, body turned with his legs out stiff.

Gerard takes out his hunting knife, and the two of us walk the two hundred feet to the motionless bodies on the hillside.

"What were you saying before you shot?" I ask him.

"Hidatsa prayer," Gerard says. "The buffalo's got a powerful spirit. I don't want him to suffer. I was asking for a clean, quick kill that will release the spirit without too much pain."

He bends down next to the larger bull, sticks his knife into its neck, and severs the jugular vein to bleed the carcass. A deep gurgle comes from the body as the soil turns dark underneath it. Gerard repeats this with the second one, and we return to the truck.

He radios the sheriffs dispatcher to say he won't need another deputy, tells his crews that the buffalo are down, and then calls his supervisor at the park's south unit to find out who will get the carcasses. A list is kept of nonprofit organizations that want the meat and are willing to pay the butchering costs. The Fort Union National Historic Site, a former trading post near the confluence of the Yellowstone and Missouri rivers, takes the heads and robes. The top of the list today includes the Jamestown Chamber of Commerce, a local Catholic church, the Fort Berthold Reservation, and the state industrial school. The supervisor says he'll contact them in that order to see who can pick up the carcasses; Gerard says he's already arranged to bring them to Watford City, where they can be ready by tomorrow.

We drive out the access road to find a county road and meet Gerard's patrol car and the other pickup. On the way, Gerard stops at a farmhouse, where he asks the rancher to bring a tractor with a front-end bucket to lift the carcasses onto the pickups, and he calls the Piggly Wiggly supermarket in Watford City to tell thc butcher he'll be bringing in two buffalo by this evening.

Our new wagon train—two pickups, a sedan, and a big tractor with an enclosed cab—returns to the buffalo. Gerard and Jimmy Olson, the head of maintenance for the park, begin work on the smaller bull. They slice circles around the bull's ankles and begin skinning the hide by pulling on the coat and cutting the fatty tissue holding the robe to the meat. It is hot, messy work done to the sounds of ripping flesh, knives being periodically sharpened on whetstones, occasional faint gurgles from the bull's windpipe, and buzzing flies. Their hands became caked with dark, dry blood. Gerard finds a third knife and gives it to me to help out. He instructs me about tugging and peeling back the hide while I slice. The skinning is harder labor than it looked.

We turn the dead bull on its back and open the chest cavity, removing the heart and putting it in a plastic garbage bag. Olson carefully slits the thin membrane near the belly, and the casing around the bull's four stomachs balloons through the opening, as if someone inside the bull had pulled the cord on an inflatable raft. Gerard, at the lower end of the abdomen, is making a cut toward Olson's, trying to be sure he doesn't rupture the full bladder. When the two cuts meet, the bags of stomachs and lungs and coils of intestines spill onto the ground.

Kristi Korb, a young park attendant from Ohio on her first day of work for the summer season, is turning a pale green as she watches us. "We always had the women do this work in the old days," Gerard tells her. "We men would go sit on the hilltop and watch for enemies."

The liver and kidneys are now exposed. Gerard slices off a piece of brownish-purple liver and offers it to the bystanders. They all decline—Kristi's answer being a wave of deeper green across her face—so he turns to me.

"You go first," he says."It'll give you some of the buffalo's strength." He watches intently as I chew the slice of liver: warm, a little crunchy, and surprisingly sweet and strong tasting. Then he eats a piece. We do the same with small slices of kidney. Gerard removes the kidneys and liver, and cuts off the bull's testicles and long penis. All but the penis are placed in the garbage bag with the heart.

Olson attaches chains to the hind legs of the carcass and has the rancher drive his tractor closer. The chains are wrapped around the tractor's bucket, and the carcass is lifted so the head is dangling just above the ground. Twisting and strenuous cutting remove the horned head. It falls in a heap with the hide, a broad, soggy, furry blanket. The bucket lifts and turns and places the skinned carcass in the bed of Olson's truck. I heft the heavy head and hide in next to it. Kristi and another attendant take the truck and first carcass to Watford City, and we start in on the older bull. It's four-thirty.

Our knives are duller, our arms are more tired, the day is hotter, and this bull is bigger than the first. Blood smears our hands, forearms, shirts, and parts of our foreheads where we have tried to wipe sweat away. We follow the same procedure—skinning the hide, opening the chest and abdomen, saving the heart, liver, kidneys, and sex organs in the garbage bag. Probably because we're trying to hurry, one cut goes too deep and the bladder breaks, spewing its contents on Olson and Gerard.

Shortly after six o'clock, we have the second carcass, the hide and head, and the heavy bag of organs in the back of Gerard's pickup truck.

Olson heads back to the park in the sedan; the rancher rumbles off in his tractor. Gerard and I take one more look at the stomachs and intestines on the hillside—"Dinner for the coyotes tonight," he says—and begin the hour's drive to the Piggly Wiggly. The pump at the well nods farewell.

The two carcasses hanging from spikes in the supermarket freezer weigh 948 pounds and 671 pounds, not counting the heads, hides, and viscera. The butcher tells us the Jamestown Chamber of Commerce and the reservation have already called, asking when they can pick up their meat. A ranger from Fort Union carts off the heads and hides. We still have the bag of organs and one of the penises in the truck bed, my "reward" for my scouting duties. Gerard has plans for them: cooking the former for tomorrow's breakfast; nailing and drying the latter on a board, until it is ready to be used as a cane.

In the Piggly Wiggly parking lot, where a small crowd gathered to watch the buffalo being unloaded, several men are still gawking at something in the back of Gerard's truck.

"What's that?" one of them asks Gerard, pointing to a three-foot-long pole of cartilage.

"What do you think it is?" he answers.

"Is it the tail?" The man doesn't want to mention what he thinks it really is.

"No," Gerard says. "What does it look like?"

"It's not that, is it?"

"You bet," Gerard says.

The man's eyes are twenty times the size of a buffalo nickel as we pull away. Gerard's blood-caked face looks a little like the profile on the coin's flip side.

I awake the, next morning in Gerard's house to the aroma of testes, kidneys, and small buffalo steaks frying in the pan. The smell is distinctive and overpowering, the same smell from the hillside where we skinned the hides.

"I'm trying to decide what that smells like," I tell Gerard, searching for descriptive images and comparisons.

He turns to me from the frying pan and fixes me with a look like Roosevelt's guide must have during their rainy hunt a hundred years ago—a look wondering if Easterners know anything about anything.

"Buffalo," he says. "It smells like buffalo."

The Buffalo Coat

By Thomas McGrath

I see him moving, in his legendary fleece,
Between the superhighway and an Algonquin stone axe;
Between the wild tribes, in their lost heat,
And the dark blizzard of my Grandfather's coat;
Cold with the outdoor cold caught in the curls,
Smelling of the world before the poll tax.

And between the new macadam and the Scalp Act
They got him by the short hair; had him clipped
Who once was wild—and all five senses wild—
Printing the wild with his hoof's inflated script
Before the times was money in the bank,
Before it was a crime to be so wild.

But history is a fact, and moves on feet
Sharper than his, toward wallows deeper than.
And the myth that covered all his moving parts,
Grandfather's time had turned into a coat;
And what kept warm then, in the true world's cold
Is old and cold in a world his death began.

Acknowledgments

Don Congdon Associates, Inc., "Velva, North Dakota: 1946 and 1964" from Not So Wild A Dream. Copyright ©1946, renewed 1964 by Eric Sevareid. Reprinted by permission of the publisher.

Richard Critchfield, "Two Views of the Depression" from Those Days. Copyright ©1968 by Richard Critchfield. Reprinted by permission of Doubleday, a division of Bantam Doubleday Dell Publishing Group, Inc. and Sterling Lord Literistic, Inc.

Dayton Duncan, "Buffalo Hunt" and "Winter Interlude" from Out West. Copyright ©1987 by Dayton Duncan. Reprinted by permission of Viking Penguin, a division of Penguin Books USA Inc.

Lucile M. Kane, June D. Holmquist and Carolyn Gilman, editors, "The Journal of Stephen H. Long, 1823" and "The Journal of James E. Colhoun, 1823" from The Northern Expeditions of Stephen H. Long: The Journals of 1817 and 1823 and Related Documents. Copyright ©1978 by the Minnesota Historical Society. Reprinted by permission of the publisher.

George T. Loher, "The Wonderful Ride" from The Wonderful Ride. Copyright ©1978 by Westword Associates. Reprinted by permission of Harper San Francisco, A Division of HarperCollins Publishers.

Richard Lyons, "Depot at Alice," "Medora," "Slant Village," "Motel," and "Fort Berthold Reservation" from Scanning the Land. Copyright ©1980 by Richard Lyons. Reprinted by permission of the author.

David Martinson, "Hardship" from Aluminum Canoe, No. 5. Copyright ©1987 by David Martinson. Reprinted by permission of the author.

Thomas McGrath, "The Buffalo Coat" from Selected Poems 1938-1988. Copyright ©1988 by Thomas McGrath. Reprinted by permission of Copper Canyon Press.

Arlene Sagness, editor, "Homestead Days" from Sods, Logs & Tar-paper. Copyright ©1985 by North Dakota Centennial Commission. Reprinted by permission of the North Dakota Extension Homemakers Council.

John Steinbeck, "A Place Unvisited" from Travels with Charley. Copyright ©1961, 1962 by The Curtis Publishing Co., ©1962 by John Steinbeck, renewed ©1990 by Elaine Steinbeck, Thom Steinbeck and John Steinbeck IV. Reprinted by permission of Viking Penguin, a division of Penguin Books USA Inc.

Solveig Zempel, editor, "An Urban Norwegian in the Red River Valley: 1883-1885" from In their Own Words: Letters from Norwegian Immigrants. Copyright ©1991 by the Regents of the University of Minnesota. Reprinted by permission of the University of Minnesota Press.